Magnetic
Field Effects on
Biological Systems

Magnetic Field Effect on Biological Systems

(based on the Proceedings of the
Biomagnetic Effects Workshop held
at Lawrence Berkeley Laboratory,
University of California, on
April 6-7, 1978)

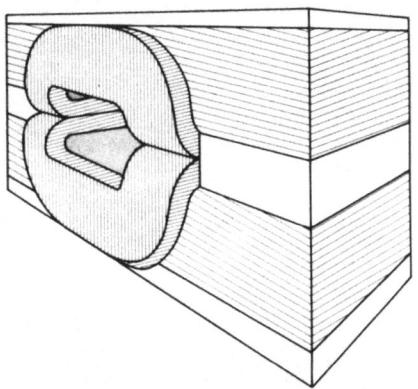

Tom S. Tenforde, Editor
Division of Biology and Medicine
Lawrence Berkeley Laboratory
University of California

Plenum Press • New York and London

Library of Congress Cataloging in Publication Data

Biomagnetic Effects Workshop, University of California, Berkeley, 1978.
  Magnetic field effect on biological systems.

   1. Magnetic fields – Physiological effect – Congresses. I. Tenforde, Tom S.
II. Lawrence Berkeley Laboratory. III. Title.
QP82.2.M3B56 1978   574.1'91   79-20739

ISBN-13: 978-1-4615-9145-0   e-ISBN-13: 978-1-4615-9143-6
DOI: 10.1007/978-1-4615-9143-6

Work supported by the Division of Biomedical and Environmental
Research, U.S. Department of Energy, under Contract W-7405-ENG-48

Acknowledgments

The staging of a Biomagnetic Effects Workshop and the production of a conference summary was a team effort on the part of several staff members at the Lawrence Berkeley Laboratory. Susan Proctor is to be especially commended for her efficient secretarial and organizational efforts; Sandy Sanford served as the conference coordinator; Deanna Fajardo handled all clerical aspects of the conference; Michael Baublitz, Gloria Chua, Ralph Dennis, Charles Hernandez, and Candace Voelker helped prepare the conference proceedings; Steve Hurst and Charles Dols prepared the newly-developed magnetic field laboratory for a tour by conference participants; Lee Davenport coordinated tour arrangements for the Bevatron and 184-inch cyclotron. The helpful advice offered by several members of the Lawrence Berkeley Laboratory scientific staff involved in biomagnetic research — especially Drs. Michael Rayborn, Cornelius Gaffey, Tracy Yang and Ruth Roots — is gratefully acknowledged. My appreciation is also extended to Dr. Edward Alpen (Director of Donner Laboratory), Dr. Cornelius Tobias (Radiation Biophysics Group Leader), and Dr. Thomas Budinger (Research Medicine Group Leader) for their encouragement and support. To these and many others go my most sincere thanks for helping to make the Workshop a successful enterprise.

Tom S. Tenforde
Editor

Contents

Chapter 1 Introduction

Tom S. Tenforde

A programmatic effort to assess the effects of magnetic field exposure on living organisms and man is under way at the Lawrence Berkeley Laboratory. This program, which is supported by the Division of Biomedical and Environmental Research of the U.S. Department of Energy, has three principal aspects. First, in a project for which I serve as the coordinator, a series of biophysical experiments are being carried out to determine magnetic field effects on molecular, cellular and whole-animal test systems. A second effort, headed by Dr. Thomas Budinger, involves epidemiological studies designed to evaluate potential health effects in groups of scientists and industrial workers who have been occupationally exposed to high magnetic fields. The third project is the establishment of magnetic field exposure guidelines by a six-member committee composed of scientists from throughout the U.S. and headed by Dr. Edward Alpen, Director of the Lawrence Berkeley Laboratory Biology and Medicine Division.

During the initial phase of this program, it became increasingly clear to all of the scientists involved that it would be a worthwhile effort to hold a Biomagnetic Effects Workshop. There were, in fact, three reasons underlying our decision to sponsor such a conference:

First of all, more than a decade has passed since there was a large conference in the United States devoted exclusively to biomagnetic research. It was therefore our intent, by staging this Workshop, to provide a forum where scientists from across the United States could hold discussions on topics of contemporary interest in the field of magnetobiology.

A second reason for holding the Workshop related to the DOE committee on magnetic field exposure guidelines, which was scheduled to meet

Tom S. Tenforde, Biology and Medicine Division, Lawrence Berkeley Laboratory, Berkeley, California 94720.

in executive session on April 7, 1978. It appeared advantageous to interface the Workshop and the committee meeting, since several of the members could then also speak at the conference.

A third reason for holding the Biomagnetic Effects Workshop was the fact that interest in magnetobiology as an independent scientific discipline is rapidly intensifying because of the large number of developing technologies that utilize high magnetic fields. It therefore seemed to us that it would be an ideal time to hold a conference to review and evaluate the current state of knowledge in magnetobiology. As an indication of the growth that has occurred in the field of magnetobiology, Figure 1 shows a histogram of the number of published papers in this field each year since the turn of the century. This compilation was based on surveys of the world literature on magnetobiology, and includes papers related both to the magnetic properties of, and magnetic field effects on, biological systems. The graph shows clearly that there has been a significant increase in activity beginning around 1960. In fact, approximately 80% of the 1800 publications in this field have appeared since 1960. But it is also interesting to note that publications over the last decade seem to have reached a plateau level of about 75 to 100 papers per year. However, another upswing might well be expected to occur during the next few years because of the current interest in studying potential biological effects resulting from exposure to the rather large fringe fields associated with magnetohydrodynamic systems, fusion reactors, magnetic energy storage facilities, and magnetically-levitated vehicles. These, of

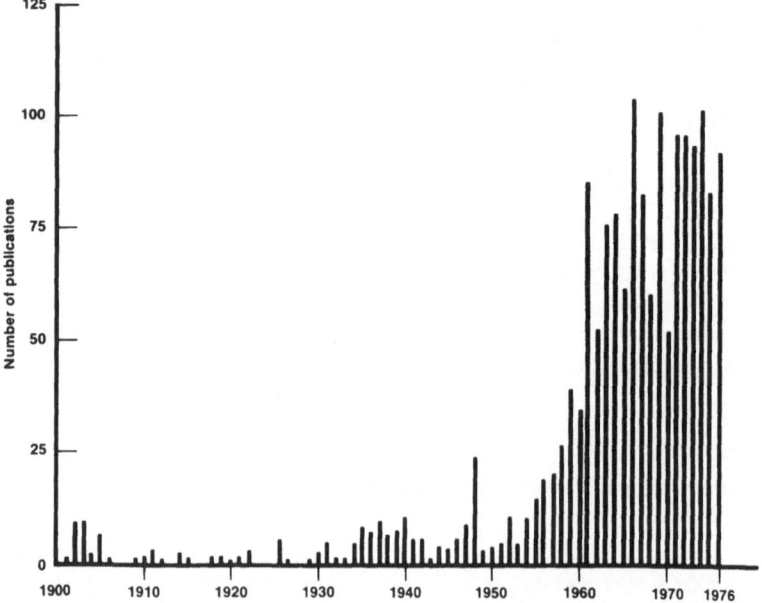

Figure 1. Yearly publications in the field of magnetobiology since 1900.

course, are only a few of the many technologies that involve high magnetic fields.

A few remarks should also be made about the choice of topics and the general organizational plan of the Workshop. In taking a broad view of the spectrum of biomagnetic effects that have been reported, it would appear that several molecular systems and lower organisms exhibit a fairly well-defined and reproducible response to magnetic fields. For example, fields of the same order of magnitude as the geomagnetic field have been shown to produce directional effects on the movement of certain birds, fish, insects, and iron-rich bacteria. However, in surveying the literature relating to organisms of greater and greater complexity, it becomes increasingly evident that well-defined magnetic field effects become fewer in number. Also, the number of controversial observations increases. This situation certainly pertains to man as well as to the mammals commonly used in laboratory experiments. The existing information in the field of magnetobiology therefore suggests that the presence of clear-cut biomagnetic effects is somehow loosely correlated with the lack of complexity of the biological test system. This view, although it is undoubtedly too simplistic, has nevertheless led to a natural grouping of subjects into the three major topics that were considered during the first day of the Workshop, namely, magnetic field effects on molecular and cellular systems, on lower organisms, and on mammals.

The second day of the Workshop was devoted to studies — both of a theoretical and experimental nature — relating to potential mechanisms of interactions of magnetic fields with biological systems. On an *a priori* basis, one might anticipate that a large number of different types of interactions could occur. To name a few, orientational effects are both expected and frequently observed in systems containing highly paramagnetic constituents, or in systems that contain diamagnetic constituents that are magnetically anisotropic. In systems with moving charge, such as the ionic currents involved in plasma membrane conductance and nerve impulse propagation, the presence of Lorentz forces and possibly Hall effect electric fields would be expected. In the presence of a time-varying field, or in systems moving rapidly through a static gradient field, induced electric fields and currents would be present. Finally, some relatively new and very interesting proposals have been made which suggest that the sensitivity of certain biological systems may result from the existence of superconducting microregions within cells, or from atomic effects such as Zeeman interactions that could influence electron transfer reactions. To some extent, it is this great diversity of potential interaction mechanisms that makes magnetobiology a complex and challenging subject. For this reason, we chose to focus during the second day of the Workshop on an in-depth consideration of these various interaction mechanisms, with a view towards providing some conceptual framework for interpreting the large number of observed biomagnetic effects.

Finally, it should be mentioned that there is also a fascinating and rapidly growing field of study involving the measurement of intrinsic, or

endogenous, magnetic fields in humans. Through the use of a superconducting magnetometer, it has become possible to detect fields at the nano-
gauss level associated with brain and cardiac activity. This topic was not
included in the Workshop program simply because, in the limited time
available, it was necessary to restrict the program to a consideration of the
effects of externally applied magnetic fields. The techniques for measuring
intrinsic magnetic fields, and the related topics of magnetoencephalography
and magnetocardiography, would in themselves form the subject matter for
a timely and interesting conference.

Chapter 2   # Magnetic Effects on Lower Organisms

Ferromagnetic Orientation in Bacteria

Richard Blakemore

Bacteria that orient and swim in a preferred direction in magnetic fields have been observed in diverse aquatic environments.[1] These magnetotactic bacteria include a variety of morphologically distinct forms. Kalmijn and Blakemore[2] found that these bacteria orient in uniform magnetic fields of about 0.5 G. Reversal of the geomagnetic field with Helmholtz coils caused the swimming bacteria to turn around in large U-turns and swim in the opposite direction. Killed bacteria also orient to align with imposed magnetic fields.

Electron microscopic examination of marine and freshwater magnetic bacteria revealed the presence of intracytoplasmic granules or "crystals" arranged in clusters or chains and contained within triple-layered membranes. Energy dispersive X-ray analysis revealed iron as the predominant elemental constituent of these particles.[1]

Blakemore and Wolfe (unpublished) recently isolated a species of magnetic bacterium from a swamp. This isolate retains its magnetism upon repeated transfer in laboratory culture media. Cells of this species contain chains of electron-dense "crystals" previously described as constituents of magnetic bacteria from mud. A non-magnetic variant of this species was cloned and subsequently examined by electron microscopy. It lacks the particles characteristic of its magnetic counterpart. Compared to magnetic

Richard Blakemore, Department of Microbiology, University of New Hampshire, Durham, New Hampshire 03824.

cells, non-magnetic variants are also less dense and produce cell suspensions much lighter in color.

Kalmijn and Blakemore[3] studied the remagnetization of swimming magnetic bacteria with the use of a critically damped RCL circuit capable of producing 0 to 600 G in the center of two Helmholtz coils. This was accomplished by discharging a 4 μF capacitor through the coils in series with a resistor. The zero-to-peak discharge occurred within 1.75 μsec, well within the time required to rotate the bacterial cells in water (several seconds). Bacteria were made to swim in a constant, uniform magnetic field ranging from zero to four times that of the local geomagnetic field. Application of a magnetic pulse of sufficient intensity and antiparallel to the direction in which the bacteria were swimming, caused the cells to turn around and swim in the opposite direction. To affect turning of 50% of a population of the predominant type from a freshwater swamp, 375 to 400 G was required. To remagnetize 50% of the predominant form from a marine marsh, 525 to 550 G was required.

Tracks of the swimming bacteria were photographed using timed exposures. The radius of the U-turn resulting from reversing the ambient (orienting) field was essentially the same as that resulting from remagnetizing the bacteria (Kalmijn and Blakemore, unpublished). Moreover, at the intensity required to remagnetize 50% of a population, no cells in the population remained completely depolarized. These results suggest that the bacteria orient through ferromagnetism and exhibit properties characteristic of single magnetic domains. No experimental evidence is presently available to suggest why the cells's magnetic movement favors northward and downward motion in the geomagnetic field.

REFERENCES

1. Blakemore, R. P., 1975. Magnetotactic bacteria. *Science* **190**:377-379.
2. Kalmijn, A. J. and Blakemore, R. P., 1977. Geomagnetic orientation in marine mud bacteria. *Proc. Intl. Union Physiol. Sci.* **13**:364.
3. Kalmijn, A. J. and Blakemore, R. P., 1978. The magnetic behavior of mud bacteria. In *Proceedings in life sciences,* eds. K. Schmidt-König and W. T. Keeton, pp. 344-345. New York: Springer-Verlag.

DISCUSSION

Q: Is the iron in magnetotactic bacteria bound to protein or some other material?

A: We do not know as yet, and in order to obtain an answer to this question, it will be necessary to separate and purify the iron-rich crystals from the bacteria. We could then determine whether these are strictly mineral substances or ferro-organic complexes.

Q: You indicated that bacteria grown in a culture medium containing everything required for optimal growth will lose the iron crystals after about six passages. Is that a genetic change or just a phenotypic change?

A: I cannot provide an answer to this question at the present time. The non-magnetic cells arose by accident and nearly obliterated my pure culture of magnetotactic bacteria. Recently, with the assistance of my wife, I have been studying cloned non-magnetic bacteria in an effort to see if they would regain a magnetic character when grown under culture conditions that are representative of the natural environment where cell growth rates are low. Some preliminary results indicate that this does occur, but further substantiation of these results is required.

Q: What is the physiological function of the iron-rich crystals?

A: There are two extreme points of view regarding this question. One view is that the crystals have adaptive significance in the behavior of the bacteria. The other view is that they are merely a byproduct of some metabolic process with no special adaptive significance. My own opinion at present would be to favor the former point of view.

Electromagnetic Guidance Systems in Fishes

Ad. J. Kalmijn

Marine sharks, skates, and rays are extremely sensitive to weak electric fields in their seawater environment.[1] Operating in the frequency range of 0 (dc) to about 8 Hz, they exhibit biologically meaningful responses to voltage gradients as low as 0.01 μV/cm.[2,3] These low-level electric fields are detected by the ampullae of Lorenzini, which comprise a delicate sensory system in the protruding snout of the elasmobranch fishes.[4] Freshwater stingrays, lower bony fish, catfish and weakly electric fish also have ampullary receptors, though of much smaller size, which render them sensitive to dc and low-frequency voltage gradients of 1 μV/cm or perhaps slightly less.[5,6] In recent years, the ampullary receptors of both marine and freshwater electrosensitive fishes have been shown to play an important role in the detection of prey and in spatial orientation.

All aquatic animals produce weak dc and low-frequency electric fields in the water that stem from potential differences at their skin-water interfaces, for example, between the mucous membranes lining the mouth and the gill epithelia in the pharynx.[5,7] Marine sharks, skates, and rays and most fresh-water electrosensitive fish take advantage of these bioelectric fields in

Ad. J. Kalmijn, Department of Biology, Woods Hole Oceanographic Institution, Woods Hole, Massachusetts 02543.

predation. When motivated by odor, they acutely zero in on the electric fields of the prey, even if it is a small animal hiding in the sand.[3] These observations, originally made under well-controlled laboratory conditions, were later verified in the field on free-roaming sharks in the ocean off Cape Cod and on wild catfish in a local freshwater pond.[8,9]

Global, wind-driven and tidal ocean currents flowing through the earth's magnetic field induce electric fields that are perpendicular to and, in the northern hemisphere, directed to the left with respect to the flow of water.[10] When measured with towed electrodes, the induced voltage gradients range from about 0.05 to 0.5 $\mu V/cm$.[11] In these fields, marine elasmobranchs may orient electrically, either to compensate for passive drift or to follow the ocean currents during migration.[5] In fresh water, the prevailing electric fields are much stronger and probably of electrochemical rather than electromagnetic origin, offering more local, territorial cues.[5] At any rate, marine elasmobranchs and freshwater electrosensitive fish have proved their ability to orient themselves in uniform electric fields in behavioral tests.[6,9]

When actively swimming through the earth's magnetic field, sharks, skates, and rays also induce local electric fields with voltage gradients that depend on the fish's compass heading.[5] As these fields are strong enough to be detected at swimming speeds of only a few centimeters per second, the elasmobranchs could very well be endowed with an electromagnetic compass sense. And indeed, sharks and rays have been observed (a) to react to non-uniformities in the geomagnetic field, (b) to rest in areas predetermined by the magnetic field direction, and (c) in training experiments, to secure food from an enclosure in the east, but not to enter a similar enclosure in the west of their tank.[12] Obviously, electromagnetic fields play an important role in the lives of these fishes.

REFERENCES

1. Dijkgraaf, S. and Kalmijn, A. J., 1962. Verhaltensversuche zur Funktion der Lorenzinischen Ampullen. *Naturwissenschaften* **49**:400.

2. Kalmijn, A. J., 1966. Electro-perception in sharks and rays. *Nature (London)* **212**:1232-1233.

3. Kalmijn, A. J., 1971. The electric sense of sharks and rays. *J. Exp. Biol.* **55**:371-383.

4. Dijkgraaf, S. and Kalmijn, A. J., 1963. Untersuchungen über die Funktion der Lorenzinischen Ampullen an Haifischen. *Z. Vergl. Physiol.* **47**:438-456.

5. Kalmijn, A. J., 1974. The detection of electric fields from inanimate and animate sources other than electric organs. In *Handbook of Sensory Physiology*, ed. A. Fessard, Vol. III/3, pp. 147-200. New York: Springer-Verlag.

6. Kalmijn, A. J., Kolba, C. A. and Kalmijn, V., 1976. Orientation of catfish *(Ictalurus nebulosus)* in strictly uniform electric fields: I. Sensitivity of response. *Bio. Bull.* **151**:415.

7. Kalmijn, A. J., 1972. Bioelectric fields in sea water and the function of the ampullae of Lorenzini in elasmobranch fishes. Scripps Institute of Oceanography Reference Series, Contribution No. 72-83, pp. 1-21. (English version of CNRS/ZWO report, 1969.)
8. Kalmijn, A. J., 1977. The electric and magnetic sense of sharks, skates, and rays. *Oceanus* **20**:45-52.
9. Kalmijn, A. J., 1978. Electric and magnetic sensory world of sharks, skates, and rays. In *Sensory biology of sharks, skates, and rays*, eds. E. S. Hodgson and R. W. Mathewson, pp. 507-528. Arlington, Virginia: Office of Naval Research.
10. Faraday, M., 1832. Experimental researches in electricity. *Philos. Trans. R. Soc. London* **122**:125-194.
11. Von Arx, W. S., 1962. *An introduction to physical oceanography*. Reading, Massachusetts: Addison-Wesley.
12. Kalmijn, A. J., 1978. Experimental evidence of geomagnetic orientation in elasmobranch fishes. In *Proceedings in life sciences*, ed. K. Schmidt-König and W. T. Keeton, pp. 348-354. New York: Springer-Verlag.

DISCUSSION

Q: Are critical fields involved in these experiments with elasmobranch organisms? That is to say, are there critical fields below which orientational cues are not observed in either the electric or magnetic field experiments? A threshold field might indicate some type of critical coupling.

A: We observe biological responses only when we remain within the normal environmental range of field strengths. When the field level drops lower and lower, the orientational response fades out gradually. Interestingly enough, in fields stronger than normal, the animals tend to ignore the stimulus altogether.

Q: From the point of view of the mechanism of signal reception, do you know as yet whether there is a direct coupling of an electric or magnetic field with the organism, or whether there is an intermediate transduction process involving convective, thermal, or acoustic processes?

A: The receptors have been studied electrophysiologically, and it has been shown that the membranes in the ampullae of Lorenzini detect the electrical stimulus directly. The reception of magnetic fields, however, probably takes place through the process of electromagnetic induction.

Q: What is the magnitude of daily and seasonal variations in the earth's electric and magnetic fields? Do sharks compensate for these variations?

A: The electric and magnetic field patterns are rather steady, but they are partly seasonal and show fluctuations. The earth's magnetic field varies only by a few percent at most; the electric fields are often much noisier, however. In our field studies on the feeding behavior of sharks, occasionally all of them show uniformly abnormal behavior. Geophysicists have found transient changes to occur in the earth's electromagnetic field of which the durations may be as short as 1-10 seconds. We will try this summer to correlate the sharks' unusual behavior with these transient fluctuations in the earth's field.

Effects of Magnetic Fields on Avian Orientation

William T. Keeton

Homing pigeons, which ordinarily rely heavily on the sun compass, can orient accurately under heavy overcast by using alternate cues that do not require time compensation.[1] The possibility that the geomagnetic field provides such a cue came from my observation that bar magnets can disrupt experienced birds' orientation under overcast skies, though not under the sun.[2] The magnets were also observed to disorient first-flight youngsters even when the sun is visible. Walcott and Green[3] were later able to change (not merely disrupt) the orientation of pigeons under overcast by attaching Helmholtz coils around the birds' heads.

Wiltschko and his colleagues have reported a rotated orientation of migratory restlessness in birds exposed to turned magnetic fields in circular test cages. They have found that the birds' sensitivity is limited to a narrow range of field strengths approximately equal to the geomagnetic field, and that the birds apparently pay no attention to field polarity but read north as that direction where the magnetic and gravity vectors form the most acute angle.[4] They have also found evidence that the birds use the earth's magnetic field to calibrate the star compass.[5,6]

Lindauer and Martin[7,8] have reported convincing evidence that the dance of scout honeybees is influenced by even minor natural fluctuations in the earth's magnetic field, suggesting a sensitivity to changes of less than 10^{-3} G and perhaps less than 10^{-5} G. Similar sensitivity has been reported for gull chicks by Southern[9] and for free-flying migrants by Moore.[10] We have found that pigeons, too, are influenced by natural magnetic fluctuations and must have a similar sensitivity.[11,12] Larkin and Sutherland[13] have found that migrants are influenced by the weak low-frequency alternating fields produced by the test antenna for the Navy's proposed Project Seafarer.

After many unfruitful attempts by others to condition birds to magnetic

William T. Keeton, Section of Neurobiology and Behavior, Cornell University, Ithaca, New York 14853.

stimuli,[14] Bookman[15] has finally succeeded in training pigeons to magnetic cues in a choice test.

In summary, there is now abundant evidence that both birds and insects are very sensitive to magnetic stimuli when they are orienting. Whether they are responsive to such stimuli under other circumstances is not yet clear.

REFERENCES

1. Keeton, W. T., 1969. Orientation by pigeons: is the sun necessary? *Science* **165**:922-928.
2. Keeton, W. T., 1971. Magnets interfere with pigeon homing. *Proc. Natl. Acad. Sci. USA* **68**:102-106.
3. Walcott, C. and Green, R. P., 1974. Orientation of homing pigeons altered by a change in the direction of an applied magnetic field. *Science* **184**:180-182.
4. Wiltschko, W. and Wiltschko, R., 1972. Magnetic compass of European robins. *Science* **176**:62-64.
5. Wiltschko, W. and Wiltschko, R., 1975. The interaction of stars and magnetic field in the orientation system of night migrating birds. I. Autumn experiments with European warblers (Gen. *Sylvia*). *Z. Tierpsychol.* **37**:337-355.
6. Wiltschko, W. and Wiltschko, R., 1975. The interaction of stars and magnetic field in the orientation system of night migrating birds. II. Spring experiments with European robins *(Erithacus rubecula). Z. Tierpsychol.* **39**:265-282.
7. Lindauer, M. and Martin, H., 1968. The earth's magnetic field affects the orientation of honeybees in the gravity field. *Z. Vergl. Physiol.* **60**:219-243.
8. Martin, H. and Lindauer, M., 1977. The effect of the earth's magnetic field on gravity orientation in the honeybee *(Apis mellifica). J. Comp. Physiol.* **122**:145-187.
9. Southern, W. E., 1972. Influence of disturbances in the earth's magnetic field on ring-billed gull orientation. *Condor* **74**:102-105.
10. Moore, F. R., 1977. Geomagnetic disturbance and the orientation of nocturnally migrating birds. *Science* **196**:682-684.
11. Keeton, W. T., Larkin, T. S., and Windsor, D. M., 1974. Normal fluctuations in the earth's magnetic field influence pigeon orientation. *J. Comp. Physiol.* **95**:95-104.
12. Larkin, T. S. and Keeton, W. T., 1976. Bar magnets mask the effect of normal magnetic disturbances on pigeon orientation. *J. Comp. Physiol.* **110**:227-232.
13. Larkin, R. P. and Sutherland, P. J., 1977. Migrating birds respond to Project Seafarer's electromagnetic field. *Science* **195**:777-779.
14. Kreithen, M. L. and Keeton, W. T., 1974. Attempts to condition homing pigeons to magnetic stimuli. *J. Comp. Physiol.* **91**:335-362.
15. Bookman, M. A., 1977. Sensitivity of the homing pigeon to an earth-strength magnetic field. *Nature* **267**:340-342.

DISCUSSION

Q: Can you speculate on how pigeons might simultaneously sense the gravity and magnetic field directions?

A: Not really, but I think that the two might be locked together in the detection process, either integratively or at the level of the sensory organ. With regard to the latter, I've been interested in looking at parts of the inner ear where gravity is detected to see if they might also be sensitive to a weak magnetic field.

Q: Presumably you could raise homing pigeons deprived of both the sun and the geomagnetic field by creating a null field or using a large field that swamps the earth's field. Could the birds then home at all, and if so, what would be the third cue that they would use?

A: We have not raised pigeons under null magnetic field conditions, but we have placed Helmholtz coils around the loft to turn the field so that magnetic north was shifted to the east. The results we got were rather ambiguous, and we haven't made sense out of them as yet. I should also correct the impression that we can completely disorient the birds by disturbing the magnetic field around them. The fact of the matter is that we can't completely make them get lost by anything we do. We can take away the sun, disturb the magnetic field around them, put frosted contact lenses on their eyes so that they can't see, plug up their noses, remove the cochlea so that they can't hear, and they still get home! It's an amazing system in that they seem to eventually correct for everything.

Q: Are there significant differences in the effects of homogeneous fields produced by Helmholtz coils and inhomogeneous fields produced by bar magnets? Also, do you concur with the findings of the German researchers that the birds sense only the direction, and not the polarity, of the magnetic field?

A: The homogeneous magnetic field produced by Helmholtz coils is certainly more effective in directing the bird's orientation. With regard to sensing the direction of the field, the important parameter appears to be the direction of the angle between the magnetic and gravity vectors. If the most acute angle between these vectors is northerly, the birds will have one orientation, and if it is southerly, they will have a reversed orientation.

Q: Are there localities where the birds are unable to orient?

A: Yes. For example, there is one particular spot located 75 miles west of our loft where pigeons always go random. We have not been able to show what is peculiar about that site, but we have determined that it is not just our pigeons. We have borrowed pigeons from breeders located in other directions from that site, and they, too, have orientational difficulties.

Effects of 60-Hz Electromagnetic Fields on Bees and Soil Arthropods

Bernard Greenberg

Soil arthropods exposed to the Navy's Wisconsin Test Facility (WTF) antenna fields (0.1 - 2.5 V/m; < 0.9 G) have been monitored annually since 1969. Their population curves and predator-prey proportions do not show any significant departures from normal fluctuations observed in control populations. Several species of soil-dwelling animals — the earthworm, slug, wood louse and redbacked salamander — were collected from sites near the WTF antenna during a three-year period (1973-1975) for measurement of metabolic rates via oxygen consumption and respiratory quotient.[1] No significant sustained differences were noted between the exposed and control populations. Comparison of metabolic rates between the summer and fall months revealed no seasonally-linked change in sensitivity to the electromagnetic fields.

Honeybees in conventional metal-containing hives under a 765-kV transmission line (7 kV/m; 10^{-3} G) showed effects not found in shielded counterparts under the line or in a control area (10 V/m; 10^{-4} G): hives failed to gain weight (p = 0.005); individual bees weighed less (p = 0.01); although egg and larval production was normal, fewer pupae were produced (p = 0.01); moisture content of honey was less (0.05 > p > 0.025); and propolization of hive entrances was extensive. Unshielded, metal-free hives under the line were normal except for some propolization of hive entrances, possibly higher hemocyte counts (p = 0.01), and possibly higher overwintering mortality. These studies are being continued.

REFERENCE

1. Greenberg, B. and Ash, N., 1976. Metabolic rates in five animal populations after prolonged exposure to weak, extremely low frequency electromagnetic fields in nature. *Radiat. Res.* **67**:252-265.

DISCUSSION

Q: Is there any suggestion of an effect related to the direction of the field?

A: Studies have been carried out where the beehives were rotated by 50 to 90 degrees relative to the overhead line. Some changes in the hives were noted, but the studies cannot be regarded as definitive.

Bernard Greenberg, Department of Biological Sciences, University of Illinois at Chicago Circle, Chicago, Illinois 60680.

Tests in the Plant *Tradescantia* for Mutagenic Effects of Strong Magnetic Fields

John W. Baum, Lloyd A. Schairer and Kenneth L. Lindahl

Biomagnetic studies at Brookhaven National Laboratory have centered on determining the possible mutagenic effects of stationary fields ranging in strength from 20 to 37,000 G. Such mutagenic effects could influence the development of germ cells, and produce somatic-cell alterations leading ultimately to carcinogenesis. For these studies, the plant *Tradescantia* was chosen because of its previously-demonstrated sensitivity to the mutagenic effects of both physical (e.g., radiation) and chemical (e.g., ozone) agents.[1-4] The endpoints for mutagenicity included (a) pink stamen hair mutation, (b) pollen abortion, and (c) micronuclei formation in the tetrad stages of pollen production.

Results obtained in eight separate trials are summarized in Table 1. A χ^2 analysis was made on each set of data. If this test indicated a difference between exposed and control specimens at the 90% confidence limit, a

Table 1. Mutagenicity tests with *Tradescantia*.

Field strength (gauss)	Exposure (days)	Difference relative to control (%)	Endpoint (score-days)	p value χ^2 Test	t Test
20-7,000	5 - 28	- -	Pinks (5-28)	- -	- -
670	6	− 0.14	Pinks (7-12)	0.014	0.07
		− 0.6	Pollen abortion (11-15)	0.11	
670	6	+ 0.2	Pinks (11-15)	0.56	
		− 1.8	Pollen abortion (18-20)	$< 10^{-5}$	0.17
1,700	6	+ 0.04	Pinks (6-17)	0.15	
8,000	6	− 0.04	Pinks (11-15)	0.5	
		− 0.2	Pollen abortion (18-21)	0.6	
9,200	0.9	+ 1.6	Micronuclei	0.07	0.5
9,400	6	+ 0.04	Pinks (7-12)	0.26	0.5
		− 0.9	Pollen abortion (7-12)	0.008	0.28
37,000	2	+ 0.03	Pinks (11-21)	0.32	
		+ 0.08	Pollen abortion (18-19)	0.89	

John W. Baum, Lloyd A. Schairer and Kenneth L. Lindahl, Safety and Environmental Protection Division, Brookhaven National Laboratory, Upton, New York 11973.

Student's *t* test was also performed. For each of the three endpoints studied, no increase in mutation rate was observed for exposed plants compared to control plants for fields up to 37 kG and for exposure periods varying from 0.9 to 6 days. The pooled data indicate that a small but not highly significant increase of 2.7 ± 2.3 (std dev) $\times 10^{-4}$ pink mutations per hair results from these magnetic field exposures.

The data presented here are used to estimate a possible mutation increase of $8 \times 10^{-5}\%$ for exposures to 200 G for 8 hours (a conventional limit for wholebody exposure at several laboratories). In comparison, mutation increases of 0.015% would be produced in this system by a permissible daily dose (20 mrem) to ionizing radiation. If one assumes that linear relations hold between field strength, exposure time, and observed effect, then these results suggest that an 8 hr/day exposure to a 40-kG field would be necessary to produce the same (or probably lower) mutation frequency as a 20-mrem X-ray exposure. Tests on other systems will, of course, be required before this can be stated as a firm conclusion. We are currently extending our mutagenicity tests to the fruitfly *Drosophila*.

REFERENCES

1. Underbrink, A. G., Schairer, L. A., and Sparrow, A. H., 1973. *Tradescantia* stamen hairs: a radiobiological test system applicable to chemical mutagenesis. In *Chemical mutagens, principles and methods for their detection*, ed. A. Hollaender, Vol. 3, pp. 171-207. New York: Plenum Press.

2. Underbrink, A. G., and Sparrow, A. H., 1974. Influence of experimental end points (dose, dose rate, neutron energy, nitrogen ions, hypoxia, chromosome volume and ploidy level) on RBE in *Tradescantia* stamen hairs and pollen. In *Biological Effects of Neutron Radiation* (Proceedings of the Symposium on the Effects of Neutron Irradiation Upon Cell Function; Munich, Germany, Oct. 22-26, 1973), pp. 185-214. Vienna: Intl. Atomic Energy Agency Publ. STI/PUB/352.

3. Schairer, L. A., Van't Hof, J., Hayes, C. G., Burton, R. M., and de Serres, F. J., 1978. Exploratory monitoring of air pollutants for mutagenicity activity with the *Tradescantia* stamen hair system. *Environmental Health Perspectives* 27:51-60.

4. Ma, T., Sparrow, A. H., Schairer, L. A., and Nauman, A. F., 1978. Effect of 1,2-dibromoethane (DBE) on meiotic chromosomes of *Tradescantia. Mutation Research* **58**: 251-258.

DISCUSSION

Q: Have you looked at end points other than mutation, for example, magnetic effects on photosynthesis or growth?

A: We have not looked specifically for such effects, but there does not appear to be an obvious influence on growth.

Q: In your tentative calculation of safety factors, you assumed that effects were linearly related to the field strength. However, some evidence for a non-linear response has been reported in studies of magnetic field effects on the metabolic rate of embryonic tissues. It would be unwise in my opinion to ignore the possibility that there may be threshold values of the field below which no effect occurs.

A: This is a very good point. There could very well be a threshold at low fields and a plateau in the response curve at high fields. However, over the range of fields that we have examined, there is no evidence for a non-linear response with the end points that we have chosen.

Q: I would like to point out that your plants are exposed to dc fields and are not moving, so that you have a completely magnetostatic situation. In the actual human occupational exposure situation, the fields are generally non-uniform and time-varying. The field may be static, but then the worker is moving to regions of differing field strength so that induced currents are established. On the basis of this consideration, don't you feel that your plant system is quite different from the occupational setting?

A: Yes, and for this reason, we are looking at other systems such as *Drosophila*. Since these organisms are moving, the exposure conditions are more typical of the human situation.

Q: What is the magnetic field direction relative to the axis of plant growth?

A: The field was vertical, and thus along the plant axis, in the electromagnet. In the permanent magnets, the field was horizontal and hence perpendicular to the axis.

Q: You are testing for somatic mutations. Have you looked for heritable genetic effects by doing experiments with replanted seeds?

A: We are not doing experiments of this nature with plants, but will be looking at heritable effects in *Drosophila*.

Chapter 3 Magnetic Field Exposure Guidelines

Edward L. Alpen

My presentation will be a sharp contrast to the discussions that dealt with lower organisms and fields with strengths of only a few gauss. Now, we are going to move up to kilogausses and talk about people. My real function here is to try to give you some insight into why there is a need to establish exposure guidelines in the immediate future, and for what particular purposes they might be suitable. The Department of Energy *ad hoc* committee which I am chairing has now completed a review of technologies that utilize magnetic fields, and I believe we have gained an understanding of the conditions for which we have to set guidelines. In the course of describing these conditions, I hope as well to give you a feeling for the range of exposures that are likely in occupational situations.

The Department of Energy committee has been given the assignment to establish interim working guidelines only. Our recommendations will not be standards by any means, nor will they be permanent. They will only propose limits that can be used as guidelines or criteria for working environments in magnetic fields.

When we were given our first statement of charge, we were asked to establish guidelines for dc fields, for dc field gradients, and lastly for ac fields. Our first step in establishing these guidelines was to hold a meeting at the Department of Energy in November 1977, where we interviewed several individuals involved in new technologies that generate large magnetic fields. One thing quickly became apparent, namely, that there appears to be no significant need or reason to establish guidelines for ac magnetic fields in the

Edward L. Alpen, Donner Laboratory, Lawrence Berkeley Laboratory, Berkeley, California 94720.

immediate future. None of the technologies that are presently developed or coming downstream in the next decade appear to have any substantial involvement with ac magnetic fields. We therefore feel that our principal problems are to work on exposure guidelines for dc fields, for pulsed dc fields, and possible dc gradients.

Let me give you some information about the characteristics of magnetic fields for which we will require guidance, particularly those arising from the new energy technologies. An interesting sidelight is that, until the last two or three years, there has been little or no interest in industrial exposure to magnetic fields. Presently, however, three Scandinavian groups are vigorously collecting occupational exposure data on what to us would be old technologies.[1-3] It has recently become quite apparent that some of these old technologies have significant magnetic fields associated with them, even though to date we have not concerned ourselves with this fact. These old technologies include induction welding, electronic heat-sealing of plastic devices, spot welders, induction furnaces, and other induction-type devices that have been in use for many decades. Associated with these types of devices are magnetic fields of up to 1000 G which are accessible to operators. More likely working environments involve exposure to tens or hundreds of gauss around these old unshielded devices.

Three important new technologies involve possible hazards from magnetic fields. The first of these technologies is the superconducting magnet energy storage system. Several of these devices are being developed throughout the nation right now. In principle, all of these devices work in the same way; they are large superconducting magnets in which energy will be stored during off-peak times. They will be very large devices containing on the order of 30 to 100 MJ of stored energy in the early prototype models, and 1000 MJ in the next generation beyond that. Table 1 shows some characteristics of three such devices, one of which is under construction at the Bonneville power plant in Washington. This is to be a 30-MJ superconducting storage

Table 1. Superconducting magnet energy storage devices.

Bonneville Prototype
Storage capacity — 30 MJ.
Floor installation. Field at the surface of the windings, 40 kG; at the dewar surface, 500 G (operator accessible).

Los Alamos Scientific Laboratory (LASL) Device
100 MJ, surface device, not buried. Field strengths not known.

Demonstration Prototype
1000 MJ, buried at a depth of 600 meters. 10 kG at the center of the buried solenoid; 200 G at the surface of the ground. A 700- to 800-meter diameter fence required for exclusion from the 200-G zone.

---

magnet, a floor installation that will be partially accessible in the working environment. The field at the surface of the magnet windings inside the cryogenic dewar is 40 kG, but operators will not have access to that area. The maximum field accessible to the operators at the dewar's surface is about 500 G, still a rather large number.

The second demonstration device will probably be built at the Los Alamos Scientific Laboratory in New Mexico. This will be a 100-MJ storage device, which is a factor of three larger than the Bonneville prototype. There do not appear to be any estimates of the magnetic fields associated with this proposed installation, but the strength will probably be of the same magnitude as the Bonneville prototype, with fields of 200 to 500 G at operator locations.

The third demonstration machine will be a 1000-MJ storage magnet. In the present design criteria, it will be built so that it has fiberglass struts to the surrounding solid rock environment, and an access gallery around the periphery of the solenoid for maintenance. At the surface of the earth, 600 m above the solenoid, the design criteria suggest that there will be approximately a 200-G field at the perimeter of an exclusion fence. A fence 700 to 800 m in diameter will be required to achieve this exclusion limit, and at the center of that zone, right over the torus, the field to which operators might be exposed could approach 500 to 800 G. In the maintenance ring there will be a field of approximately 20-kG strength during operation, but I think that one of two options will be necessary there: either the storage device will have to be emptied of energy before access is permitted, or the individuals entering will have to wear protective shielding. For all intents and purposes, the maximum fields that will be accessible to man in the superconducting magnet energy storage devices will be on the order of 200 to 500 G.

The second energy technology that the DOE Committee has considered, the prototype fusion reactor, is the one that has brought to the fore the increasing interest in magnetic fields. It was originally thought that the fusion reactor people were the ones who would require ac magnetic field exposure guidelines. At this point, it is clear that they have no such need. Table 2 summarizes the state of knowledge of magnetic field characteristics for present and near-term experimental fusion reactors, including the Princeton Tokamak which will be the first large-scale reactor to be completed. With pulsed reactors under normal operating conditions, there will be a 2- to 60-G field in zones accessible to operators. If these devices are operated steady-state, the normal operation levels would be on the order of 100 to 300 G. In the control rooms for fusion reactors of any type, it will be necessary to sharply reduce these magnetic fields for reasons other than operator safety. They will be reduced to 1-2 G or less by shielding in order to enable operation of oscilloscopes and other electronic devices. Consequently, the field will be less than 1 G where most of the operators are located. The nature of these fields is such that there will be pulses, ranging from a tenth of a millisecond up to 5000 seconds. The pulse duration is a measure of the engineering success of the system, since the longer the pulse, the more efficient the

Table 2. Magnetic field characteristics of fusion reactors (near-term experimental devices).

A. Machines designed for pulsed mode	Normal operation	Control failure
Magnetic field level (G)	2 - 60	100 - 10,000
Magnetic field gradient (G/m)	~50	100 - 20,000
Pulse duration (msec)	0.1 - 5000	0.1 - 5,000
Pulse frequency (day^{-1})	20 - 200	n.a.
Operator exposure (day^{-1})	~15 min	a few pulses
B. Machines designed for steady state mode	**Normal operation**	**Control failure**
Magnetic field level (G)	100 - 300	1000
Magnetic field gradient (G/m)	~200	200 - 10,000
Operator exposure (day^{-1})	~8 hr	few min

device. If plasma stability is ultimately achieved, the reactors will be dc machines. I think that most scientists in the fusion reactor business feel that dc operation is not likely to be achieved for at least several decades after pulsed operation is initiated. In the interim, it is likely that there will be pulse devices operating at frequencies of 20 to 200 pulses per day with a pulse duration of a couple of hundred milliseconds.

This is the present status of experimental fusion reactors. If we move forward to the planned facilities that might be available in the late 1980's or beyond (Table 3), all of these will be very much larger devices operating at the 1000- to 1500-MW level. There will still be potential exposure to magnetic fields during normal operations, but these fields will generally be less than 100 G in strength. However, these devices will have rather sharp magnetic field gradients. Data given in Table 3 for the control-failure mode states the maximum possible exposure to people who will be in maintenance and operating spaces outside the control room. These fields can rise as high as 10 kG with very large gradients, and the duration could be quite long — on the order of 30 seconds. Again, there appears to be no need for any ac field guidelines for fusion operations as we currently see them.

The third technology that the DOE Committee has considered is the whole field of accelerator operations, including bubble chambers that involve particularly large magnetic fields. Figure 1 shows stray magnetic field plots around the large Brookhaven bubble chamber. The black dot shows the typical position and exposure level for the individual who must go near the chamber several times a day during an 8-hour shift to change film cassettes. The dot represents his head. In the procedure of changing film cassettes, the field to which he is exposed could range from a high of 15 kG to a low of 6 kG for approximately 8 to 10 minutes. These are very high field levels, and they will be present every time the person goes in for a film-pack change.

Table 3. Magnetic field characteristics of fusion reactors (prototypes, late 1980's and beyond).

A. Machines designed for pulsed mode	Normal operation	Control failure
Magnetic field level (G)	20	10,000
Magnetic field gradient (G/m)	~500	5,000
Pulse duration (msec)	30,000	30,000
Pulse frequency (day^{-1})	~100	n.a.
Exposure time (day^{-1})	~1 hr	several pulses

B. Machines designed for steady state mode	Normal operation	Control failure
Magnetic field level (G)	5	3000
Magnetic field gradient (G/m)	1	100
Exposure time (day^{-1})	~8 hr	less than 1 hr

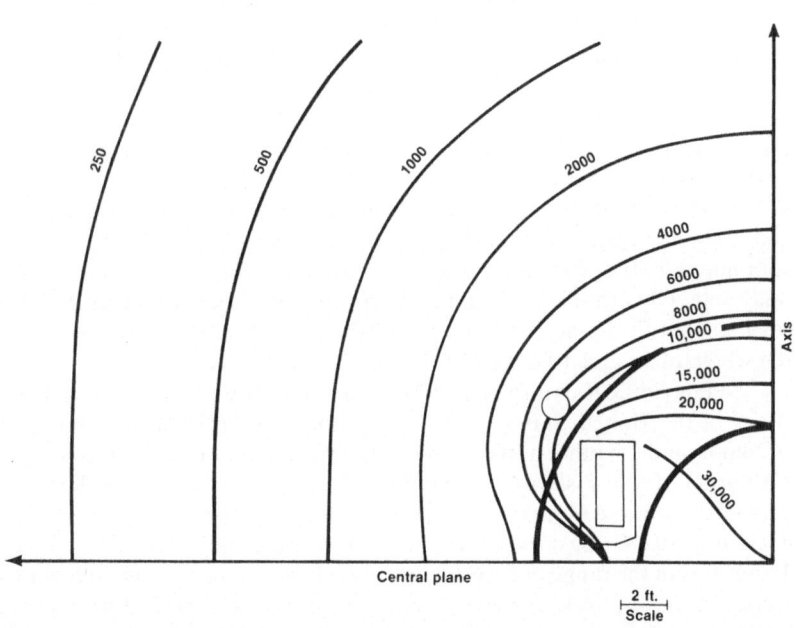

Figure 1. Plot of magnetic field strengths (in gauss) around the hydrogen bubble chamber at Brookhaven National Laboratory in Upton, New York. The dot represents the location of an operator during the procedure of changing film cassettes.

Now where are we as far as standards for operations are concerned today? Are there any existing criteria? It appears that there is no official magnetic field exposure criteria extant anywhere in the world. There are proposed standards from several sources, and I would like to discuss these to provide an idea of what they are relative to the exposures that might be experienced in newly developing technologies. Table 4 shows an unofficial standard recommended by Dr. A. M. Vyalov in the U.S.S.R. Within the Soviet Union, these standards have no status at all. They have not been officially accepted,

Table 4. Regulation of magnetic field exposure (USSR),
(Unofficial) Standards recommended by Vyalov (see Refs. 4 and 5).

Homogeneous fields (dc)		Gradient fields (dc)	
Whole body	300 G	Whole body	500 - 2000 G/m
Hands	700 G	Hands	1000 - 2000 G/m

and have not even been any more than mentioned by Vyalov. By private communication, however, I have been made aware that these guidelines are being used in some of the high-energy physics establishments in Russia as unofficial guidelines for occupational exposure at accelerators, bubble chambers, and similar devices. As you can see, the guidelines are only for dc fields, but they do have one unique characteristic, namely, that they set a standard both for homogeneous fields and for gradient fields. They also are intended for an 8-hour continuous workday. These are only occupational standards, and are not for the general public. Dr. Vyalov has suggested that during a continuous workday, exposures should be no higher than 300 G to the whole body and 700 G to the hands alone. These are rather large numbers. He also suggests that field gradients should not be more than 500 to 2000 G/m over the whole body and 1000 to 2000 G/m over the hands.

Magnetic field exposure standards are currently in use in two laboratories in the United States – one at the National Accelerator Laboratory in Illinois and the other at the Stanford Linear Accelerator Center (SLAC) in California. Table 5 shows the unofficial National Accelerator Laboratory standard. It stipulates that an individual who will be exposed to a field exceeding 10 kG must have the approval of the Senior Radiation Officer. Exposures in the range of 5 to 10 kG to the whole body are acceptable for up to one hour, but these exposure conditions cannot be exceeded without the permission of the Senior Radiation Officer. With fields in the range of 100 G to 5 kG, work in the exposure zone is required to be "minimized." I might remark that these National Accelerator Laboratory guidelines permit operators to carry out bubble chamber film changes since they are exposed to fields in excess of 10 kG for only 15 minutes at a maximum.

Table 6 presents the Stanford Linear Accelerator standards, which are

Table 5. Regulation of magnetic field exposure (USA),
          (Unofficial) National Accelerator Laboratory standards.

10 kG for any time
Only with permission of Senior Radiation Officer.

5 - 10 kG
Whole body up to one hour. Can be exceeded with permission of Senior Radiation Officer.

100 G - 5kG
Work in area is minimized.

also unofficial. These have been used at SLAC since it was first established eight years ago. They suggest, not unlike the proposed Russian criteria, that exposure to dc fields for extended periods should not exceed 200 G to the whole body or the head, or 2000 G to the arms and hands. For short periods, these limits are increased, respectively, to 2000 and 20,000 G. This latter number again permits film changing at the SLAC bubble chamber.

Table 6. Regulation of magnetic field exposure (USA),
          (Unofficial) Stanford Linear Accelerator standards.

Whole body or head (dc)		Arms and hands (dc)	
Extended periods	200 G	Extended periods	2000 G
Short periods (minutes)	2000 G	Short periods	20,000 G

In conclusion, the current unofficial standards contain a wide range of numbers, but they only differ by about an order of magnitude, which isn't really bad in this business.

REFERENCES

1. Galliano, P. G., 1977. Measurements of strong electromagnetic fields in industrial environments. Abstract J2, *International Symposium on the Biological Effects of Electromagnetic Waves, Airlie, Va., Oct. 30 - Nov. 4.*
2. Hietanen, M., Kalliomaki, P. L., Lindfors, P., Ristila, J., and Kalliomaki, K., 1977. Measurement of electric and magnetic field strengths near industrial radiofrequency heaters. Abstract J3, *International Symposium on the Biological Effects of Electromagnetic Waves, Airlie, Va., Oct. 30 - Nov. 4.*

3. Mild, K. H., and Olsson, G., 1977. Preliminary studies of electrical and magnetic fields at 27 MHz: field studies and response of blood cells. Abstract J4, *International Symposium on the Biological Effects of Electromagnetic Waves, Airlie, Va., Oct. 30-Nov. 4.*

4. Vyalov, A. M., 1967. Magnetic fields as a factor in an industrial environment. *Vestnik* **8**:52-58.

5. Vyalov, A. M., 1971. Clinico-hygienic and experimental data on the effects of magnetic fields under industrial conditions. In *Influence of magnetic fields on biological objects*, ed. Y. Kholodov, National Technical Information Sevice Rept. JPRS 63038 (1974).

DISCUSSION

Q: You remarked that in the foreseeable future there is no need to consider standards for ac fields. In the recent past, we have heard about the microwave surveillance techniques being used in Russia on our embassy people. These fields seem to have some biological effects, so vis-a-vis that, don't you think we should worry about ac fields also?

A: I am talking about only magnetic fields, and those are microwave fields.

Q: Are there any standards for microwave fields?

A: There are official standards for radiofrequency radiation in Russia, Czechoslovakia, Poland, and America.

Q: From my own personal experience, I would suggest that you might give some special consideration to sensitive populations in gradient fields. I am referring, for example, to people with metal clips in their heads who would experience discomfort in a magnetic field.

A: There are two sensitive populations that must be excluded from magnetic fields of any significant magnitude — people wearing cardiac pacemakers and individuals with metallic prostheses. They should not be there. I feel that there is simply no excuse for people with metallic prostheses or pacemakers to work in a magnetic field.

Q: Have you considered magnetic fields associated with superconducting transmission of electricity?

A: That technology lies so far in the future that we do not have to be concerned with it as yet. The only new thing that is being examined in electric power transport is refrigerated transmission. This method is entirely different from superconducting transmission. The refrigerator is in a prototype stage, and there is no magnetic field associated with it except in the rotaries and compressors for the refrigerant.

Chapter 4 Magnetic Effects on Mammals

Magnetic Field Interactions in Man and Other Mammals: An Overview

Asher R. Sheppard

Of the many experimental reports concerning small laboratory mammals exposed to magnetic fields, few give reliable evidence on the extent and character of biomagnetic interactions.[1-4] Overall, the lack of conclusive evidence and a wide variation in the field strength, frequency, and exposure duration used in different studies have resulted in a confused picture.

Reports of magnetic field effects cluster about effects on blood composition, alterations in normal physiologic parameters of growth and activity, effects on the immune system, altered neutral function and altered animal behavior. For example, Odintsov[5] reports reduced phagocytic activity and reduced leukocyte counts in mice exposed to a 200-G field at 50 Hz (6.5 hr for 15 d), and Toroptsev[6] found histopathologic changes in lung, testes, ovaries, kidney, liver, bone marrow, etc. using similar fields. Changes in plasma hormones are reported from studies with rats exposed to a 200-G, 50-Hz field.[7] Barnothy[4] reports on the reduction of white cells in mice exposed to a 4-kG stationary field. Squirrel monkeys exposed to a 200-G steady field (4 hr/d for 10 d) showed increased levels of urinary steroids.[8] These and similar reports concerning such labile physiologic parameters point to the need for further investigations.

With regard to neural effects, the EEG was reported to be altered in rabbits exposed to a 200-G stationary field[3] and, in monkeys exposed to very

Asher R. Sheppard, Jerry L. Pettis Memorial V.A. Hospital, Loma Linda, California 92357.

high fields (20 to 90 kG), large effects on EEG amplitude and power spectrum were seen.[9] Vagal inhibition in frog hearts exposed to a strong (4 kG) steady field was attributed to a change in acetylcholinesterase activity.[10] Öberg[11] reports neural stimulation by magnetic fields.

Interest in possible effects on the central nervous system by weak fields is stimulated by the evidence that birds possess a magnetic compass,[12,13] by several surprising findings concerning bird orientation in very weak electromagnetic fields,[14,15] and by the extraordinary electric field sensitivity of certain fish.[16] The effect of weak, low frequency electric fields on calcium binding in neural tissue provides a possible neurochemical basis for these influences.[17,18]

In a pilot study, Beischer and coworkers[19] observed elevated serum triglyceride levels in humans exposed to a 1-G field at 45 Hz (for 22.5 hr). Doubts about this finding must be emphasized because no such effect was seen in a better-controlled study in monkeys.[20]

A study by deLorge with monkeys indicates that fields of about 10 G (mostly at 45 Hz) have no noticeable effect on behavior,[21] and Schmitt and Tucker found that human subjects do not perceive 10-G, 60-Hz fields.[22]

In contrast, it is clear that humans in a strong magnetic field ($>$ 100 G) experience excitation of the visual pathway, the "magnetic phosphene." Relatively little is understood of the mechanism by which the sensation of flickering, shimmering light patterns is produced. Present evidence supports the notion of an effect at the retina rather than at cortical structures, but direct cortical stimulation may play a role under some circumstances (and, for example, direct electric stimulation of the cortex produces phosphenes). Following d'Arsonval's initial description of the effect it was determined that phosphenes may be generated by (a) alternating fields in the range of about 10 to 100 Hz or, (b) by the transients associated with energizing or de-energizing a large magnet, or (c) by motion of the subject's head in a steady field. The phenomenon is strongly dependent on the head's orientation in the field; magnetic stimuli are effective at frequencies well above the flicker fusion frequency.[23-27]

Assuming that phosphenes are generated in the retina, two hypotheses are immediately available: (a) there is a magnetic field effect on the photosensitive molecules of the photoreceptors resulting in neural stimulation via the photochemical pathway or, (b) the induction of eddy currents in the retina directly affects retinal neurons that are already specialized for the processing of graded potential changes. Research into these questions is under way at the Lawrence Berkeley Laboratory in the U.S. and in Sweden.[28] The Swedish scientists observed spike trains in retinal ganglion cells following exposure to a magnetic field and they have characterized a frequency-dependent threshold for phosphenes in humans.[29] Surprisingly, when phosphenes are produced under certain ambient light conditions the threshold above 30 Hz differs in subjects with abnormal color vision, although the localization of phosphenes in the peripheral visual field suggests little involvement for the color-sensitive cone receptors of the foveal region.

REFERENCES

1. Schiff, A., 1978. A quantitative review of human susceptibility to magnetic fields. Lawrence Livermore Laboratory, Livermore, California, Rept. UCID-17773.
2. Sheppard, A. R. and Eisenbud, M., 1977. *Biologic effects of electric and magnetic fields of extremely low frequency.* New York: New York University Press.
3. Kholodov, Y. A., 1971. *Influence of magnetic fields on biological objects.* National Technical Information Service, Rept. JPRS 63038 (1974).
4. Barnothy, M. F., ed., 1964 and 1969. *Biological effects of magnetic fields.* Volumes 1 and 2. New York: Plenum Press.
5. Odintsov, Y. N., 1965. *The effect of a magnetic field on the natural resistance of white mice to Listeria infection.* National Technical Information Service, Rept. JPRS 62865 (1974).
6. Toroptsev, I. V., et al., 1971. Pathologoanatomic characteristics of changes in experimental animals under the influence of magnetic fields. In *Influence of magnetic fields on biological objects,*ed. Y. Kholodov, National Technical Information Service, Rept. JPRS 63038 (1974).
7. Udintsev, I. V. and Moroz, V. V., 1974. Response of the pituitary-adrenal system to the action of a variable magnetic field. *Bull. Exp. Biol. Med.* **77**:641-642.
8. Friedman, H. and Carey, R. J., 1972. Biomagnetic stressor effects in primates. *Physiol. Behav.* **9**:171-173.
9. Beischer, D. E. and Knepton, J. C., 1966. The electroencephalogram of the squirrel monkey *(Saimiri sciureus)* in a very high magnetic field. Naval Aerospace Medical Institute, Pensacola, Florida, Rept. NAMI-972.
10. Young, W. and Gofman, J. W., 1965. Magnetic fields, vagal inhibition and acetylcholinesterase activity. Lawrence Livermore Laboratory, Livermore, California, Rept. UCRL-12389.
11. Öberg, P. A., 1973. Magnetic stimulation of nerve tissue. *Med. Biol. Eng.* **11**:55-64.
12. Walcott, C. and Green, R., 1974. Orientation of homing pigeons altered by a change in the direction of an applied magnetic field. *Science* **184**:180-182.
13. Keeton, W. T., 1971. Magnets interfere with pigeon homing. *Proc. Natl. Acad. Sci. USA* **68**:102-106.
14. Southern, W. E., 1975. Orientation of gull chicks exposed to Project Sanguine's electromagnetic field. *Science* **189**:143-145.
15. Larkin, R. P. and Sutherland, P. J., 1977. Migrating birds respond to Project Seafarer's electromagnetic field. *Science* **195**:777-779.
16. Kalmijn, A. J., 1966. Electro-perception in sharks and rays. *Nature (London)* **212**:1232-1233.
17. Bawin, S. M. and Adey, W. R., 1976. Sensitivity of calcium binding in cerebral tissue to weak environmental electric fields oscillating at low frequency. *Proc. Natl. Acad. Sci. USA* **73**:1999-2003.
18. Blackman, C. F., Elder, J. A., Weil, C. M., Benane, S. G., and Eichinger,

D. C., 1977. Two paramet rs affecting radiation-induced calcium efflux from brain tissue. Abstract M-2, *International Symposium on the Biological Effects of Electromagnetic Waves, Airlie, Va., Oct. 30-Nov. 4.*

19. Beischer, D. E., Grissett, J. D., and Mitchell, R. E., 1973. Exposure of man to magnetic fields alternating at extremely low frequency. Naval Aerospace Medical Research Laboratory, Pensacola, Florida, Rept. NAMRL-1180.

20. Grissett, J. D., Kupper, J. L., Brown, R. J., and Kessler, M. J., 1977. Data supplement to Interim Research Report, June, 1976. Naval Aerospace Medical Research Laboratory, Pensacola, Florida.

21. de Lorge, J., 1974. A psychobiological study of rhesus monkeys exposed to extremely low frequency-low intensity magnetic fields. Naval Aerospace Medical Research Lab., Pensacola, Fla., Rept. NAMRL-1203. (Available from NTIS as AD 000078.)

22. National Academy of Sciences, 1977. Biologic effects of electric and magnetic fields associated with proposed Project Seafarer. Report of the Committee on Biosphere Effects of Extremely-Low-Frequency Radiation, National Research Council, Washington, D.C.

23. Magnusson, C. E. and Stevens, H. C., 1911. Visual sensations caused by changes in the strength of a magnetic field. *Am. J. Physiol.* **29**:124-136.

24. Dunlap, K., 1911. Visual sensations from the alternating magnetic field. *Science* **33**:68-71.

25. Barlow, H. D., Kohn, H. I., and Walsh, E. G., 1947. Visual sensations aroused by magnetic fields. *Am. J. Physiol.* **148**:372-375.

26. Seidel, D., 1968. Der Existenzbereich elektrisch und magnetisch induktiv angeregter subjektiver Lichterscheinungen (Phosphene) in Abhangigkeit van ausseren Reizparametern. *Elecktromedizin* **13**:208-211.

27. Oster, G., 1970. Phosphenes. *Sci. Am.* **222**:82-87.

28. Lövsund, P., Öberg, P. A., and Nilsson, S. E. G., 1977. A method for the study of retinal ganglion-cell activity induced by ELF magnetic fields. Abstract K-1, *International Symposium on the Biological Effects of Electromagnetic Waves, Airlie, Va., Oct. 30 - Nov. 4.*

29. Lövsund, P., Öberg, P. A., and Nilsson, S. E. G., 1977. Quantitative determination of threshold values of magnetophosphenes. Abstract K-2, *International Symposium on the Biological Effects of Electromagnetic Waves, Airlie, Va., Oct. 30 - Nov. 4.*

DISCUSSION

Q: As I recall, many of the early phosphene experiments suffered from the deficiency that the experimenters were also the subjects. There were generally no clear descriptions of subjective test procedures. Is this still the case?

A: Yes, many of phosphene experiments were conducted in the manner you described, which would certainly lead to a bias on the part of the subject observing the phosphenes.

Q: In the dc field experiments with squirrel monkeys, what was the effect on the brain frequency distribution?

A: It was shifted towards higher frequencies in the range 20 to 40 Hz, with a peak at 30 Hz. These are usually quiet frequencies in the EEG, which under normal conditions shows peaks at 8 and 12 Hz.

Effects of Magnetic Fields on Behavior in Nonhuman Primates

John de Lorge

Largely through the impetus of Dr. Dietrich Beischer, research on the behavioral effects of magnetic fields in monkeys has been a continuing theme at our laboratory during the last two decades. One of the first studies was conducted by John S. Thach in which three squirrel monkeys, *Saimiri sciureus*, were conditioned to respond on a visual vigilance task and subsequently exposed to dc magnetic fields in the core of a water-cooled Bitter magnet.[1] Response was greatly suppressed by fields of 70 kG or more and a threshold seemed to exist between 46 and 70 kG. A second experiment in a superconducting magnet in which eight squirrel monkeys were trained on several operant tasks found similar suppressive effects up to 97 kG. In addition, two of the monkeys regurgitated when exposed to these higher fields. All of these effects were found to be reproducible.

Recent studies by the present investigator using rhesus, *Macaca mulatta*, and squirrel monkeys have revealed no behavioral effects of extremely low frequency (ELF) electromagnetic fields.[2-5] Magnetic fields from 3 to 10 G and electric fields from 1 to 29 V/m (rms) at frequencies of 7, 10, 15, 45, 60 and 75 Hz were used. No consistent effects, other than one unreplicated effect on general activity, were observed. Such operant performance as reaction time, inter-response time, matching-to-sample and overall lever responding were not consistently influenced by the applied fields. Occasionally, one animal's performance on one task would indicate a statistically significant difference between exposed and control conditions, but the effect generally could not be replicated in the other animals or in the same animal at a different time. It is concluded that reported effects of low magnetic fields on behavior are probably related to other uncontrolled environmental variables.

John de Lorge, Naval Aerospace Medical Research Laboratory, Pensacola, Florida 32508.

REFERENCES

1. Thach, J. S., 1968. A behavioral effect of intense dc electromagnetic fields. In *Use of Nonhuman Primates in Drug Evaluation*, ed. H. Vagtborg, pp. 347-356. Austin: University of Texas Press.

2. de Lorge, J., 1972. Operant behavior of rhesus monkeys in the presence of extremely low frequency-low intensity magnetic and electric fields: Experiment 1. Naval Aerospace Medical Research Lab., Pensacola, Fla., Rept. NAMRL-1155. (Available from NTIS as AD 754058.)

3. de Lorge, J., 1973a. Operant behavior of rhesus monkeys in the presence of extremely low frequency-low intensity magnetic and electric fields: Experiment 2. Naval Aerospace Medical Research Lab., Pensacola, Fla., Rept. NAMRL-1179. (Available from NTIS as AD 764532.)

4. de Lorge, J., 1973b. Operant behavior of rhesus monkeys in the presence of extremely low frequency-low intensity magnetic and electric fields: Experiment 3. Naval Aerospace Medical Research Lab., Pensacola, Fla., Rept. NAMRL-1196. (Available from NTIS as AD 774106.)

5. de Lorge, J., 1974. A psychobiological study of rhesus monkeys exposed to extremely low frequency-low intensity magnetic fields. Naval Aerospace Medical Research Lab., Pensacola, Fla., Rept. NAMRL-1203. (Available from NTIS as AD 000078.)

DISCUSSION

Q: I am concerned that the animals used in your experiments may have been perturbed in their behavioral responses because of *ad lib* feeding and confinement. The *ad lib* condition could cause a loss of the integrated circadian rhythm and affect behavioral responses. Do you feel that this factor could affect performance in an operant conditioning test regardless of the magnetic field exposure?

A: We have been running comparison studies with isolated control monkeys. Telemetry data show beautiful circadian rhythms similar to those for free-running animals. It appears that as the animals adapt to their environmental conditions, there is no evidence of a stress response.

Magnetic Field Effects on Rodents

Gabriel G. Nahas

Rats exposed to homogeneous magnetic fields produced by permanent magnets and ranging in strength from 200 to 1200 G were examined for possible vascular or histopathologic effects.[1] Capillary circulation in the mesoappendix was observed by an *in vivo* microscopic technique after exposure to 500 G for 6, 12 and 30 days. Rats exposed to 200 to 1200 G for 29-32 days were sacrificed for analysis of blood chemistry, body and organ weights, and histopathological examination of vascular tissues and 10 selected organs. The *in vivo* study of capillary circulation and the histopathologic results showed no hemodynamic alterations or intravascular thrombosis associated with magnetic field exposure. Hematology revealed no changes in hematocrit, white blood cell count, hemoglobin level or coagulation time in the exposed rats. Except for a nonpathologic congestion of the spleen, no histopathological effects were noted in any organs following exposure to the field. One unexpected observation was an increase in body and organ weights of young rats exposed to magnetic fields relative to matched controls. As the number of animals studied was small, and the factors controlling growth are numerous, further studies will be required before drawing conclusions on the significance of increased growth rates observed in animals exposed to magnetic fields of such intensity.

REFERENCE

1. Nahas, G. G. Boccalon, H., Berryer, P., and Wagner, B., 1975. Effects in rodents of a one-month exposure to magnetic fields (200-1200 gauss). *Aviat. Space Environ. Med.* **46**:1161-1163.

DISCUSSION

Q: The increase in brain weight of the exposed rats seemed surprisingly large. How old were your rats at the time of exposure?

A: They were 6 to 8 weeks old, and doubled in weight during the period of exposure.

Q: Did you note any behavioral effects in the exposed animals?

A: We did not observe the animals during the magnetic field exposure.

Gabriel G. Nahas, Department of Anesthesiology and Pathology, College of Physicians and Surgeons, Columbia University, New York, New York 10032.

Q: Did you control light-dark cycles?

A: Yes, there were ports in the exposure chamber that allowed light to penetrate. An electronic timer was used to provide a 12-hr light, 12-hr dark cycle.

Q: Could the atypical weight gain be related to hormonal effects resulting from the magnetic field exposure?

A: In order to approach that question, one could measure the weights of hypophysectomized rats exposed to a magnetic field. However, this experiment has not yet been done.

Studies on Biomagnetic Effects in Mice

Max W. Biggs

The growth rate of young mice, hematologic parameters in adult mice, and the growth of Erlich ascites tumors were examined in a magnetic field ranging in strength from 8,800 to 14,400 G.[1] The experiments were designed to test earlier reports by the Barnothys and their colleagues of positive magnetic field effects in each of these systems.[2,3] In two separate experiments, the growth rate of male mice weighing 12 g at the start of the experiment was unaffected by a continuous 15-day exposure to the magnetic field. In addition, no growth effects were observed following termination of the exposure. In adult mice, no alternations were observed relative to matched controls in body weight, white blood cell count, differential blood count, hematocrit, red blood cell count, or liver and spleen weights following magnetic field exposure for periods of 16-23 days. Ascites tumor growth was similarly unaffected by the presence of the magnetic field, based on both the rate of tumor growth and the mean survival time of tumor-bearing hosts.

REFERENCES

1. Eiselein, B. S., Boutell, H. M., and Biggs, M. W., 1961. Biological effects of magnetic fields — negative results. *Aerosp. Med.* **32**:383-386.
2. Barnothy, M. F. (ed.), 1964. *Biological effects of magnetic fields.* Vol. 1. New York: Plenum Press.
3. Barnothy, J. M., Barnothy, M. F., and Boszormenyi-Nagy, I., 1956. Influence of a magnetic field upon the leukocytes of the mouse. *Nature (London)* **181**:1785-1786.

Max W. Biggs, Department of Industrial Medicine, Lawrence Livermore Laboratory, Livermore, California 94450.

DISCUSSION

Q: It appeared from your data that the mean tumor cell count was consistently lower in the exposed group than in the control group. Was this generally the case?

A: In the experiments that we reported, there was a small difference but it was not statistically significant.

Chapter 5   Magnetic Effects in Cellular and Molecular Systems

Mechanisms of Magnetic Field Interactions With Retinal Rods

Felix T. Hong

Chalazonitis and co-workers[1] reported in 1970 that isolated frog rod outer segments in aqueous suspension can be oriented by a homogeneous magnetic field of 10 kG. The equilibrium orientation is parallel to the applied field. Furthermore, the two ends of a rod appear equivalent in a magnetic field. This latter observation suggested that the effect is either paramagnetic or diamagnetic. In either case, the effect can be due to (a) magnetic anisotropy, (b) "form" anisotropy, or (c) inhomogeneity of the applied field. The second and third mechanisms are ruled out, because the corresponding estimated magnetic potential (orientation) energy is not large enough to overcome thermal fluctuation. Numerical estimation based on the mechanism of magnetic anisotropy indicates that it is impossible to orient individual molecules in a rod with a field strength of 10 kG. However, two major molecular constituents, visual pigment rhodopsin and phospholipid, are oriented along the axial direction in a rod. If either molecule possesses a small anisotropy, the anisotropy will be additively summed in a rod and increased by a factor of 3×10^9 (rhodopsin) or 10^{13} (phospholipid). The crucial parameter is the *summed anisotropy,* which is the sum of the anisotropy of all the individual oriented anisotropic molecules, $\sum_i V_i \Delta \chi_i$, where $\Delta \chi_i$ and V_i are the anisotropy of the volume susceptibility and the total effective volume of species i, respectively. An elementary calculation

Felix T. Hong, Department of Physiology, Wayne State University, School of Medicine, Detroit, Michigan 48201.

using classical magnetic theory leads to expressions that correctly describe the time course of orientation.[2-4] Becker et al.[5] and Chabre[6] have confirmed the effect and have shown that the anisotropy is due mainly to the protein moiety of rhodopsin (possibly an α-helical region) rather than phospholipid or the chromophore retinal. From the measurement records reported by Chagneux and Chalazonitis,[7] the summed anisotropy in a retinal rod is $(1.7 \pm 0.2) \times 10^{-18}$ cm^3. This summed anisotropy, if attributed to rhodopsin alone, yields an anisotropy of volume susceptibilities ($\chi_{axial} - \chi_{radial}$) of 1.2×10^{-8} c.g.s. unit.[4] On illumination, a decrease of this anisotropy by 20% and 9% has been found by Chagneux et al.[8] and by Chabre,[6] respectively. The magneto-orientation effect is a manifestation of the extraordinary order in the rod membrane structure, and has been utilized as an experimental tool in obtaining oriented biological samples for the purpose of X-ray crystallography[9] and neutron diffraction studies,[10] in addition to fluorescence depolarization and linear dichroism studies.

Similar magnetic orientation mechanisms might play a physiological role in other systems. Accumulating evidence has shown that some birds use the terrestrial magnetic field as a cue in orientation and navigation. Experiments carried out by Wiltschko and Wiltschko[11] on European robins suggested that the magnetic compass of the bird can only detect the axial direction of the magnetic field, and cannot differentiate the north from the south. Instead, the bird derives the sense of north from interpreting the acute angle between the inclination of the axial direction of the magnetic field and the direction of the gravity vector. A similar experiment carried out in South America would be a good check of this theory. Thus, birds trained in the northern hemisphere will perhaps be confused in the southern hemisphere. If this turns out to be the case, then ordered biological structures (such as *pecten oculi*) instead of ferromagnetic particles should be examined in the search for the mysterious magnetic compass.

REFERENCES

1. Chalazonitis, N., Chagneux, R., and Arvanitaki, A., 1970. Rotation des segments externes des photorécepteurs dans le champ magnétique constant. *C. R. Acad. Sci. Ser. D.* **271**:130-133.

2. Hong, F. T., Mauzerall, D., and Mauro, A., 1971. Magnetic anisotropy and the orientation of retinal rods in a homogeneous magnetic field. *Proc. Natl. Acad. Sci. USA* **68**:1283-1285.

3. Hong, F. T., 1973. On the importance of being ordered. II. Diamagnetic anisotropy in ordered biological structures. Ph.D. dissertation, The Rockefeller University, New York, Appendix III:157-182.

4. Hong, F. T., 1977. Photoelectric and magneto-orientation effects in pigmented biological membranes. *J. Colloid Interface Sci.* **58**:471-497.

5. Becker, J. F., Trentacosti, F., and Geacintov, N. E., 1978. A linear dichroism study of the orientation of aromatic protein residues in

magnetically oriented bovine rod outer segments. *Photochem. Photobiol.* **27**:51-54.

6. Chabre, M., 1978. Diamagnetic anisotropy and orientation of helix in frog rhodopsin and meta II intermediate. (Submitted for publication.)

7. Chagneux, R. and Chalazonitis, N., 1972. Evaluation de l'anisotropie magnétique des cellules multimembranaires dans un champ magnétique constant (segments externes des bâtonnets de la rétine de grenouille). *C. R. Acad. Sci. Ser. D.* **274**:317-320.

8. Chagneux, R., Chagneux, H., and Chalazonitis, N., 1977. Decrease in magnetic anisotropy of external segments of the retinal rods after a total photolysis. *Biophys. J.* **18**:125-127.

9. Chabre, M., 1975. X-ray diffraction studies of retinal rods. I. Structure of the disc membrane, effect of illumination. *Biochim. Biophys. Acta* **382**:322-335.

10. Chabre, M., Saibil, H., and Worcester, D. L., 1975. Neutron diffraction studies of oriented retinal rods. *Brookhaven Symp. Biol.* **27**(III),77-85.

11. Wiltschko, W. and Wiltschko, R., 1972. Magnetic compass of European robins. *Science* **176**:62-64.

DISCUSSION

Q: What are the relative contributions of membrane phospholipids and rhodopsin to the magnetic anisotropy of the retinal rods?

A: If we assume that the anisotropy of phospholipid is similar to that of a long chain fatty acid such as stearic acid [which was reported by K. Lonsdale in *Proc. R. Soc. London* **A171**:541 (1939)], a retinal rod should orient perpendicular to the applied magnetic field, instead of parallel. In other words, the summed anisotropy of phospholipid partially cancels the summed anisotropy of rhodopsin in a retinal rod. Therefore, the anisotropy of rhodopsin might be larger than the number I cited (1.2×10^{-8} c.g.s. unit).

Orientation of Biological Membranes and Cells in Magnetic Fields

Nicholas E. Geacintov

Photosynthetic particles, such as chloroplasts isolated from spinach leaves, algal cells, etc., suspended in aqueous solutions, can be oriented by applying an external magnetic field of 10 kG or more.[1] This orientation can be observed by monitoring the anisotropic optical properties (linear dichroism, fluorescence and light scattering) and photoelectric properties of aqueous suspensions of these oriented particles.[2]

The physical basis of these effects can be adequately explained in terms of an anisotropy in the diamagnetic susceptibility of at least some of the oriented molecular components of the membranes. Furthermore, the particles must possess an intrinsic symmetry, for example, in spinach chloroplasts the membranes are not randomly oriented with respect to each other, but are stacked on top of one another (structures called "grana").

In order to identify such orientation effects when they occur in biology, it is necessary to consider the physical basis of this phenomenon. It arises because of an interaction between the magnetic field H and the anisotropic diamagnetic susceptibility of oriented groups of molecules within the membranes. If we designate the volume of such domains of oriented molecules by V, the magnetic susceptibility tensor by $\overline{\overline{X}}$ with orthogonal components X_1, X_2, X_3, and the magnetic energy by U, it can be shown that:

$$U = -1/2 \iiint \vec{H} \cdot \overline{\overline{X}} \cdot \vec{H} \, dV. \tag{1}$$

A macroscopic orientation of particles occurs because U is significantly larger than kT for certain orientations of the particles. If we assume that $|X_3| \gg |X_2| \approx |X_1|$, then

$$U = 1/2 \, VH^2 \left\{ X_1 + \Delta X \cos^2\theta \right\}, \tag{2}$$

where the anisotropy $\Delta X = X_3 - X_1$, and θ is the angle between X_3 and H. The direction X_3 is fixed with respect to the anisotropic optical properties of the particles (e.g., the optical absorption); experimentally, the degree of orientation is monitored by observing how the magnitude of the linear dichroism signal, for example, varies with H. When further increases in H no longer produce any changes in the linear dichroism, the orientation is assumed to be complete; experimentally, these orientation curves (linear dichroism plotted as a function of H) are sigmoidal in shape. By calculating the Boltzmann average of $<\cos^2\theta>$, utilizing eq. (1) and the function

Nicholas E. Geacintov, Department of Chemistry, New York University, New York, New York 10003.

exp [-U/kT], a theoretical orientation function ($\cos^2\theta$ vs. H) is obtained which is also sigmoidal in shape. In fact, experimentally determined and theoretically calculated orientation curves superimpose on one another quite well, providing support for the proposed orientation mechanism. Other evidence for this mechanical orientation effect can be obtained by varying the viscosity of the suspension, and by monitoring the relaxation of the orientation (transition from the orientated to a randomly oriented state) after the magnetic field is removed.

For intact *Chlorella* cells, complete orientation is obtained when the parameter α is ~4-5. This parameter is defined by $\alpha^2 = 1/2 \, \Delta XVH^2/kT$ and the magnetic field strength is about 10 kG. The magnetic energy U for complete orientation is ~16-25 times larger than kT at ambient temperatures.

The observation of magnetic orientation effects depends critically on the product $V \cdot \Delta X$, which is found to be 10^{-20} cm^3 for whole *Chlorella* cells. The smaller ΔX, the larger the size of the domain containing the oriented molecules that is necessary to achieve an observable orientation. Utilizing readily available magnetic fields of 10-20 kG, we have found that the minimum size of the biological particles that can be oriented by magnetic fields is of the order of one micron. However, smaller particles such as small protein complexes can be readily oriented by electric fields. Also, while a magnetic field of 15 kG is necessary to completely orient spinach chloroplasts, these particles are easily oriented in the presence of weak alternating electric fields (60 Hz) as low as 20-30 V/cm.[3] It is thus evident that the orienting magnetic forces are much weaker for normally available magnetic fields than for commonly available electric fields.

In summary, orientation forces in biological systems must be taken into account in assessing any possible biological effects of magnetic fields.

REFERENCES

1. Geacintov, N. E., Van Nostrand, F., Becker, J. F., and Tinkel, J. B., 1972. Magnetic field induced orientation of photosynthetic systems. *Biochim. Biophys. Acta* **267**:65-79.
2. Becker, J. F., Geacintov, N. E., and Swenberg, C. E., 1978. Photovoltages in suspensions of magnetically oriented chloroplasts. *Biochim. Biophys. Acta* **503**:545-554.
3. Gagliano, A. G., Geacintov, N. E., and Breton, J., 1977. Orientation and linear dichroism of chloroplasts and sub-chloroplast fragments oriented in an electric field. *Biochim. Biophys. Acta* **461**:460-474.

DISCUSSION

Q: There is a real question of whether structural proteins are anchored in the chloroplast membrane and affect the amount of energy required for magnetically induced reorientation. These oriented proteins could also affect the relaxation time. Do you feel that one could explain the orientation of the entire chloroplast in a magnetic field solely by means of membrane protein reorientation?

A: When the chloroplasts are broken, the smaller subchloroplast fragments do not display any orientation (or linear dichroism) in magnetic fields of ~15 kG. Since at these field strengths the whole chloroplasts are completely oriented, we believe that the broken chloroplast experiments show that there is no reorientation of any of the molecular components *within* the membranes.

The oriented proteins, however, may significantly contribute to the diamagnetic anisotropy of whole chloroplasts, and thus to their orientation in magnetic fields.

Q: Regarding the magnetic orientation of intact *Chlorella,* there exist chlorophyll-deficient mutants that have tubular photosynthetic units. Do you feel that it would be of value to examine the orientation of these mutant cells?

A: Such orientation experiments may be useful in determining whether there is an overall anisotropy in the arrangement of the tubular membranes within the whole cells. If they are randomly oriented, the cells will not orient in a magnetic field.

Enzyme-Substrate Reactions in High Magnetic Fields

Mitchell Weissbluth

Three sets of experiments were performed[1-3] with the objective of disclosing magnetic field effects, if any, on the rates of enzyme-substrate reactions. In the first series, the systems RNA-ribonuclease and succinate-cytochrome c reductase were investigated in magnetic fields up to 48 kG with exposure times of 5-6 minutes. The second series was conducted at the National Magnet Laboratory at the Massachusetts Institute of Technology. It consisted of the systems RNA-ribonuclease, horseradish peroxidase (which catalyzes the oxidation of o-dianisidine by hydrogen peroxide), tyrosinase-L-tyrosine, aldolase-FDP (fructose 1, 6-diphosphate) exposed to magnetic fields in the range of 85-170 kG with exposure times of 2-20 minutes. The peroxidase was additionally placed in the fringe field at 85 kG so as to provide a field gradient with a variation of about 30% over the volume of the sample. In the final series, also at the National Magnet Laboratory, trypsin-

Mitchell Weissbluth, Department of Applied Physics, Stanford University, Stanford, California 94305.

BAPA was exposed to 220 kG for 9 minutes and finally, the enzyme alone was "pretreated" for 65-220 minutes at 208 kG. In all cases, the reaction rates were independent of the magnetic field.

A survey of the literature indicates that in some instances positive results have been obtained. The most recent example is the work of Komolova et al.[4] who observed no effect of a field of 3200 G on RNAase but an increase of 30% in the reaction rate of DNAase. Further work in this area would appear to be indicated.

REFERENCES

1. Maling, J. E., Weissbluth, M., and Jacobs, E. E., 1965. Enzyme-substrate reactions in high magnetic fields. *Biophys. J.* **5**:767-776.
2. Rabinovitch, B., Maling, J. E., and Weissbluth, M., 1967. Enzyme-substrate reactions in very high magnetic fields. I. *Biophys. J.* **7**:187-204.
3. Rabinovitch, B., Maling, J. E., and Weissbluth, M., 1967. Enzyme-substrate reactions in very high magnetic fields. II. *Biophys. J.* **7**:319-327.
4. Komolova, G. S., Erygin, G. D., Vasileva, T. B., and Egorov, I. A., 1972. Effect of a high strength constant magnetic field on enzymatic hydrolysis of nucleic acids. *Dokl. Akad. Nauk SSSR Ser. Biol.* **204**:995-997.

DISCUSSION

Q: How closely was the temperature regulated during your assays of enzyme activity?

A: We were generally able to keep the temperature variation withing 0.5° C.

Q: Did you look for effects on enzyme function of low magnetic fields, say, less than 50 G?

A: In one series of experiments, the magnetic fields ranged from 0 to 48 kG, and no effect on enzyme function was observed in either the low or high field regions.

Q: In previous studies where magnetic fields were observed to affect enzyme reaction rates, do you know if the exposures were performed at temperatures close to a phase transition point? If so, could this be a cause of the conflict between your results and those of other workers?

A: Our experiments were carried out at temperatures close to 25° C. As far as I recall, most of the experimental results reported by others were obtained at similar temperatures.

Effects on Cell Function Resulting From Exposure to Strong Magnetic Fields at 4° K

George I. Malinin, William D. Gregory, Luigi Morelli, and Paul S. Ebert

At ambient temperature, it is difficult to obtain unambiguous quantitative results of magnetic field effects on measurable cellular functions. These difficulties may be due to a number of physical and metabolic factors, for example, the difference between kT and the energy of the applied magnetic field, the induction of EMF's by the motion of particulate cellular components and, finally, by the metabolic asynchrony of target cells. We have therefore exposed cells to magnetic fields at 4° K, thus transforming a fluid cellular target into a solid state system. In all experiments, target cells were frozen to 4° K in the presence of an appropriate cryoprotective agent, exposed to a given magnetic field for a predetermined time, thawed rapidly and allowed to resume metabolism under optimal growth conditions. Cell cultures manipulated in exactly the same manner, but not exposed to a magnetic field, served as controls. The preliminary results of these experiments may be summarized as follows:

Morphological Transformation of Target Cells: Murine heteroploid L-929 and human diploid WI-38 fibroblasts exposed to a 5-kG magnetic field at 4° K subsequently develop morphologically distinct cell types which can be propagated from generation to generation.[1]

Inhibition of DNA Synthesis in Murine Lymphocytes: DNA synthesis in spleen-derived murine lymphocytes was measured by ^{3}H-thymidine incorporation. ^{3}H-TdR incorporation in concanavalin A stimulated lymphocytes previously exposed to a 40-kG field at 4° K was drastically reduced as compared to controls.

Suppression of Hemoglobin Synthesis in Friend Erythroleukemia Cells: Hemoglobin synthesis in murine erythroleukemia cells could not be induced with butyrate following exposure of the cells to a 40-kG field at 4° K. In this series of experiments, control cultures showed a normal response to butyrate induction as measured with the benzidine assay.

These data seem to indicate that the exposure of target cells to strong magnetic fields at 4° K results subsequently in reproducible and measurable alterations of cellular metabolic indicators. Some preliminary measurements of this magnetization of the cells with a superconducting SQUID magnetometer indicate a possible difference in cell magnetization between control and target cells.

George I. Malinin, William D. Gregory, and Luigi Morelli, Department of Physics, Georgetown University, Washington, D.C. 20057.

Paul S. Ebert, Virus Tumor Biochemistry Section, Laboratory of DNA Tumor Viruses, National Cancer Institute, Bethesda, Maryland 20014.

REFERENCE

1. Malinin, G. I., Gregory, W. D., Morelli, L., Sharma, V. K., and Houck, J. C., 1976. Evidence of morphological and physiological transformation of mammalian cells by strong magnetic fields. *Science* **194**:844-846.

DISCUSSION

Q: In the study of cell magnetization, were the susceptibility measurements carried out on cells in the frozen state, or had they been thawed?

A: The magnetic susceptibility was determined for frozen cells. Those that had been exposed to a magnetic field appeared to have some paramagnetic character, whereas the control cells did not.

Q: Is the temperature at which you expose cells to the magnetic field critical for the effects that you have observed? Have you done experiments at liquid nitrogen rather than liquid helium temperature?

A: We have some preliminary evidence to indicate that liquid helium temperature is required, but the results are not as yet completely clear.

Q: Would you care to speculate on the mechanism underlying the cellular effects of magnetic fields applied to cells in the frozen state? Are the effects attributable to genetic alterations, or possibly to extranuclear cell components such as membranes and mitochondria?

A: It is difficult to say what the mechanism is. The reduced sensitivity of exposed lymphocytes to mitogen stimulation could result from an inhibition of DNA synthesis or from membrane effects.

Q: Have you done clonogenic assays to determine the reproductive capacity of the exposed cells relative to controls? What is the effect of the magnetic field on the lifespan of the W I-38 fibroblasts?

A: We did not look into these questions since we lack the culture facilities required for long-term studies.

Q: Is there any possibility that your results may be attributable to an induced electric field?

A: We were quite concerned about that possibility, and in our experiments with an electromagnet, we were careful to increase and decrease the field very slowly to minimize the induced EMF. Our experiments with a superconducting magnet were carried out while the magnet was operating in a persistent mode. and there did not appear to be any induced EMF.

Effects of a Transverse Magnetic Field on the Dose Distribution of High Energy Electrons and on the Responses of Mammalian Cells *in vitro* to X-Rays

Ravinder Nath, Sara Rockwell, Paul Bongiorni and Robert J. Schulz

High-energy electron beams (3-45 MeV) have received widespread clinical application in radiotherapy. As opposed to high-energy X-rays, the electrons have a limited depth of penetration, which is a function of electron energy. When a transverse magnetic field is applied to a dosimetry phantom, it can be predicted that an incident beam of high-energy electrons will be made to follow a spiral path in the course of slowing down. The effect of this spiralling is to enhance the dose at depth for the same entrance dose. Monte Carlo calculations of this phenomenon for 70-MeV electrons and 60-kG fields have resulted in dose distributions similar to those produced by high linear energy transfer radiations such as negative π mesons. Experiments at Yale University with 50- and 55-MeV electrons traversing a 20.5-kG field support the theoretical predictions.[1]

A potential complication of using the technique of magnetic dose enhancement is that the patient must be placed in a high magnetic field. The possibility of changes in the radiation responses of the tumor and normal tissues, and of hazards to the patient from exposure to intense magnetic fields, must be thoroughly examined. We have begun by examining the effects of magnetic fields on mammalian cells *in vitro*. Exposure of unirradiated EMT6 mouse mammary tumor cells to an almost uniform magnetic field of 1400 G for up to 48 hr did not alter the proliferation or viability of the cells.[2] Exposure of cells to this magnetic field during irradiation with 120-kV X-rays did not alter the repair of sublethal damage. Exposure of plateau phase cultures to the magnetic field after irradiation for times up to 24 hr did not alter the repair of potentially lethal damage. An almost uniform magnetic field of 5700 G did not alter the repair of sublethal or potentially lethal damage of x-irradiated EMT6 cells and did not alter the viability of unirradiated cells during exposures of up to 6 hr.

Most recent experiments were performed with a different cell line (DON Chinese hamster cells) and higher field strengths. Exposure of these cells to an almost uniform field of 20.5 kG or to a non-uniform field of 17.4 kG with a gradient of 2300 G/cm for 2 hr did not alter their viability. The survival curves for cells exposed to 30 MV X-rays delivered in single treatment or in two treatments separated by an incubation period of 2 hr outside the magnetic field were not affected by the almost uniform 20.5-kG magnetic field.

So far, no significant changes in the viability or the radiation responses

Ravinder Nath, Sara Rockwell, Paul Bongiorni and Robert J. Schulz, Department of Therapeutic Radiology, Yale University, School of Medicine, New Haven, Connecticut 06510.

of mammalian cells *in vitro* have been observed in cells exposed to high magnetic fields.

REFERENCES

1. Nath, R. and Schulz, R. J., 1978. Modification of electron beam dose distributions by transverse magnetic fields. *Medical Physics* **5**:226-230.
2. Rockwell, Sara, 1977. Influence of a 1400 gauss magnetic field on the radiosensitivity and recovery of EMT6 cells *in vitro*. *Int. J. Radiat. Biol.* **31**:153-160.

DISCUSSION

Q: Do you know why the Monte Carlo calculation of the depth-dose distribution did not completely agree with the experimentally measured profile?

A: The Monte Carlo calculation by Shih was for 70-MeV electrons and a field strength of 60 kG. Our measurements were carried out with 50- and 55-MeV electrons and a 20-kG field. Therefore, the parameters were different, but the theoretical predictions are qualitatively in agreement with our experimental measurements.

Q: What were your control conditions? Did you use a dummy magnet, or an electromagnet with the field turned off?

A: In our experiments with permanent magnets, the control cells were placed either in a dummy magnet (with a field of less than 10 G), or in the same incubator as the magnet, but outside of the field. In the electromagnet experiments, the control cells were placed in the magnet with the field off. I might mention that in all cases we monitored the temperature closely, since a small change in temperature could affect the radiation response.

Q: In the electromagnetic control experiments, did you determine what the field is when the magnet is turned off?

A: The remanent field is on the order of a few gauss.

Q: Have you considered using nervous tissues since these may be sensitive to magnetic fields? For example, do you plan to do experiments with cultured neuroblastoma cells?

A: No. Our experiments have been carried out only with cultured mouse mammary tumor cells and Chinese hamster cells. We also plan to extend our studies to the *in vivo* situation in order to determine the radiation response of intact tumors in the presence of a magnetic field.

Effect of Magnetic Fields on the Drug-Induced Contractility of the Ciliate *Spirostomum*

Earl Ettienne, Angela Ripamonti and Richard B. Frankel

The effect of a homogeneous magnetic field on the mechanism of cellular contraction has been studied in the ciliate *Spirostomum ambiguium*. In these experiments, ciliate contraction was induced by addition of 10-μM PDS (2,2'-dipyridyldisulfide), a sulfhydyl oxidizing agent that acts on the microsomal respiratory chain to reduce cytoplasmic ratios of NADPH/NADP and GSH/GSSH.[1] These ratios have been implicated in the oscillatory regulation of ionized free calcium concentration in the cytoplasm over ranges that include threshold levels for *Spirostomum* contraction.[2,3] It has also been demonstrated that the ciliate contractility is fundamentally like that of striated muscle and that there is a Ca^{++} dependent regulation of contractile elements.[4]

Spirostomum were exposed to 9200 G, and the number of PDS-induced contractions per minute were monitored over an interval of 20 min in populations of cells exhibiting synchronous contractions. At 7, 10, 13 and 16 min following incubation in PDS, the values for peak contraction frequencies were depressed in the exposed samples relative to controls by 50%, 37%, 33%, and 33% respectively. Based on a Student's t test, all of these differences were significant at the level $p < 0.001$. The duration of the relaxation phase in a magnetic field was also observed to be extended over controls.

Marked effects of the magnetic field were also observed on cell mortality after stimulation with PDS. Table 1 summarizes data on percentage survivors at 10, 20, 30 and 40 min after PDS stimulation in populations of control cells and in cells exposed to 5000 and 9200 G. With both field strengths, a significant decrease in cell survival was observed over the entire incubation period.

The above results indicate that an imposed magnetic field acts to decrease contraction frequency and to increase the duration of the contraction cycle in *Spirostomum*. Based on the known physiology of *Spirostomum*, these observations suggest that the presence of a magnetic field serves to decrease the enzymatic transport of Ca^{++} out of the cytoplasm following contraction. We are presently pursuing experiments with ^{45}Ca to determine whether the locus of the magnetic field effect is intracellular Ca^{++} transport membranes.

Earl Ettienne, Department of Physiology, University of Massachusetts Medical School, Worcester, Massachusetts 01605.

Angela Ripamonti and Richard B. Frankel, Francis Bitter National Magnet Laboratory, Massachusetts Institute of Technology, Cambridge, Massachusetts 02139.

Table 1. Survival of *Spirostomum* following simultaneous stimulation with PDS and exposure to a magnetic field.

t (min)	% **Survival ± 1 std dev**	**Significance***
	Zero Field : n = 15	
10	99 ± 3.22	
20	93.7 ± 9.6	
30	74.1 ± 10.2	
40	44.2 ± 27.8	
	5000 G : n = 15	
10	91.6 ± 9.96	$p < 0.01$
20	76.12 ± 5.92	$p < 0.01$
30	44.9 ± 15.3	$p < 0.0005$
40	7.56 ± 8.5	$p < 0.0005$
	9200 G : n = 15	
10	89.68 ± 12.2	$p < 0.001$
20	66.3 ± 13.6	$p < 0.001$
30	33.12 ± 16	$p < 0.0005$
40	4.02 ± 6.3	$p < 0.0005$

*Significance of difference in % survival relative to controls.

REFERENCES

1. Levine, M. A. and Ettienne, E., 1978. Coenzyme-dependent suppression of 2,2-PDS induced contractions in *Spirostomum*. *Microbios Letters* **20**:81-93.
2. Ettienne, E. and Dikstein, S., 1974. Contractility in *Spirostomum*

provides for nonelectrogenic calcium regulation through energy-dissipative metabolic processes in the absence of membrane excitability. *Nature (London)* **250**:782-784.

3. Dikstein, S. and Hawkes, R. B., 1976. Metabolically regulated cyclical contractions in microinjected *Spirostomum:* a pharmacological study. *Experientia* **32**:1029-1031.

4. Ettienne, E. M., 1970. Control of contractility in *Spirostomum* by dissociated calcium ions. *J. Gen. Physiol.* **56**:168-179.

DISCUSSION

Q: Did you closely regulate the temperature in your experiments, since this could drastically affect the rate of calcium transport?

A: Yes, the temperature of both control and exposed organisms was maintained at 21 ± 1°C.

Chapter 6   Long-Range Electromagnetic
Field Interactions at Brain
Cell Surfaces

W. Ross Adey

My initial plan was to present a rather simple view of the ways in which brain cells might sense fields that are intrinsic to brain tissue as well as fields that are imposed in the environment. I became convinced, however, that some speculation relating to cellular and molecular mechanisms underlying the perception of electromagnetic fields would also be appropriate. Therefore, I will try to present some facts, or near-facts, and then discuss models of ways in which interaction with environmental fields may occur in the central nervous system.

It is currently a very exciting time in cerebral neurophysiology, because many of the classic concepts that most of us learned in undergraduate biology are being reexamined. Specifically, the ways in which brain cells communicate with each other, and in many instances the preclusion of impulses as a basis for this communication, are ideas that I will try to explain. There is, for example, strong evidence for a system of communication between brain cells that uses only slow-wave processes rather than nerve impulses. Second, there is evidence that brain cells sense weak electric fields in their environment, far below the electric gradients characterizing synaptic processes. Third, there is virtually conclusive evidence that the interaction of electromagnetic fields in brain tissues involves non-equilibrium processes — phenomena which are new to the biologist accustomed to models of the Hodgkin-Huxley type. I think the last concept, that interactions may occur through non-equilibrium processes, has opened some very remarkable doors

W. Ross Adey, Research Service 151, Jerry L. Pettis Memorial V.A. Hospital, Loma Linda, California 92357.

in our thinking, at least about the modeling of cell-to-cell communication in brain tissue.

At the outset, I shall try to bridge any communication problems that may exist between those of us who are primarily in the field of biology and those who are from the physical sciences. We can begin with the construction of nervous systems, using *Hydra* as an example of a very simple living organism. The *Hydra* has sensory cells in its body wall that are connected to a muscle coat lying in the deeper layer of the body wall. These cells communicate in a direct way with each other; effectively, the system is one in which a sensory stimulus produces a direct response in the motor apparatus. That is a very far cry from what evolved in the invertebrates, and then in vertebrates. In the vertebrates, including mammals, there is a system of interneurons between the sensory receptors and the motor apparatus. In effect, we can consider the central nervous system of higher animals as a gross overgrowth of the interneuronal system. What I will be addressing in considering cerebral function in its most simplistic form is this interneuronal function, namely, those cells that are interposed between the sensing apparatus and the motor apparatus for the purposes of transacting and storing information in the context of learning and remembering.

How is the brain neuron different from other nerve cells and how will this bear on our discussion of sensing field effects? Briefly, a cerebral neuron has a very small cell body (typically about 10 μm in diameter, though a few go to 30 or 40 μm), but a huge branching dendritic apparatus which has, until quite recently, baffled us conceptually. Clues from many directions point to the dendrites of cerebral cells as the place where one sees the most characteristic function of brain tissue, as for example in the growth of the cortex from birth. At birth, cells in the cortex appear like the ragged trunks of trees after a forest fire, with the cell bodies dispersed quite widely and apparently separate within the cortex. At four months the cell bodies are no closer together and no more numerous, but sprouting out from them are dendritic branches. By 15 months, the system has grown to a very complex network of dendrites which we now know to make functional contacts with each other. As a consequence, much of the intrinsic communication in the cortex occurs through dendro-dendritic contacts, and not by axons sending impulses. The dendro-dendritic concept has occasioned a great deal of thought and exercised many groups interested in central nervous organization. In a recent review, Schmitt, Dev and Smith[1] described the organization of nervous tissue organization in the retina and in the olfactory bulb. Horizontal cells in the retina make contact with other parts of the system through dendrites, and the granule cells of the olfactory bulb are also activated by dendro-dendritic connections. These contacts were observed in both the retina and the olfactory bulb, and it is now believed that this dendro-dendritic organization is widespread through the cortex as a whole. The electric processes occurring between dendrites are in general not impulses, but rather a slow electrical activity passing across synaptoid junctions that are often reciprocal and allow activity to pass in both directions.

Given that dendrites are the prime sources of this slow electric activity in the cortex, we may observe an integral of this activity as the EEG. Figure 1 show an intracellular record where the cell has a membrane potential of about 50 mV.[2] The baseline in the membrane potential is constantly perturbed by waves that resemble the EEG recorded in the fluid around the cell; but whereas the EEG is only a few microvolts in amplitude, the large waves inside the cell, coming from its dendrites, are on the order of 10 mV or more. This signal is thus about 200 times larger than the small part of the activity that leaks into the fluid around the cell. The records shown in Figure 1 demonstrate that when a cat is monitored while asleep, the waves are large and slow inside the cell and in the EEG. When the animal is awake, the intracellular wave is small and fast and so is the EEG. One can observe that from time to time the cell fires. That is a high-order transform of the wave process going on inside the cell unceasingly. It leads to the question of whether or not the component of the intracellular wave appearing in the fluid around the cell, which we call the EEG, may also have an informational content. Do brain cells sense the EEG? That is a very fundamental question which, to this time, has tended to be answered in the negative with the additional statement that the EEG is nothing more than the noise of the brain's motor. Therefore, if fields around the head entered the domain of the EEG, they would be expected to be as ineffective as the EEG in changing the animal's behavior.

Let me also emphasize that if one makes a spectral analysis on both the EEG and the neuronal wave as shown in Figure 2, one observes a very similar

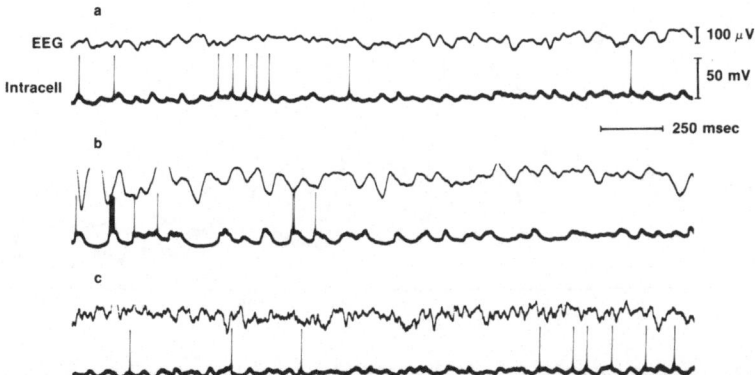

Figure 1. Simultaneous records from cortical surface (EEG) and an intracellularly placed glass micropipette (INTRACELL) from a cat while asleep (a and b) and awake (c). The intracellular waves are approximately 200 times the amplitude of the extracellular EEG (note calibrations), and are generated continuously. Action spikes occur in the intracellular records at infrequent intervals on the depolarizing peaks of neuronal waves. (From Elul, R., 1972, *Int. Rev. Neurobiol.* **15**:227-272; reproduced by courtesy of Academic Press.)

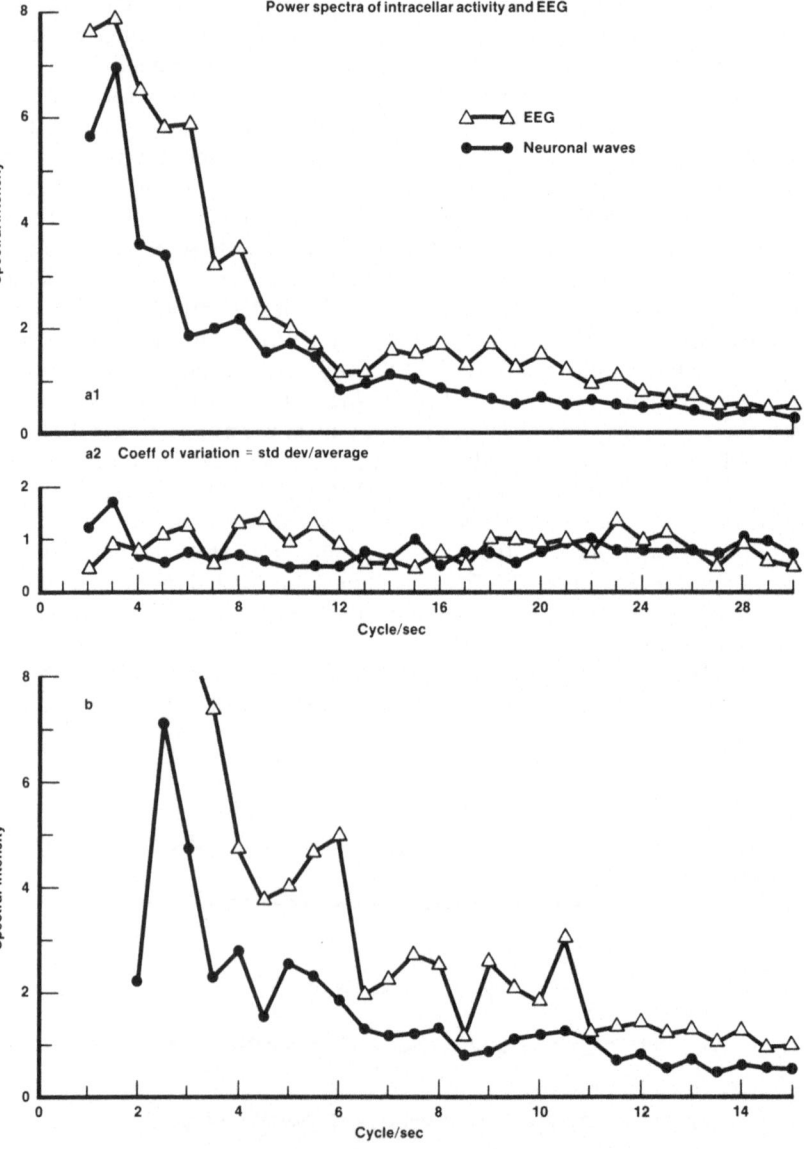

Figure 2. Spectral analyses of simultaneous EEG and neuronal wave trains, showing the similarity of the spectral contours. (From Elul, R., 1968, Data acquisition and processing in biology and medicine. In *Proceedings of the 1966 Rochester Conference* 5:93-115; reproduced by courtesy of Pergamon Press.)

contour, frequency-by-frequency, between the EEG and the neuronal wave. It should also be emphasized that the neuronal wave recorded inside the cell is non-synchronous with the EEG; it is noncoherent with the EEG in the same domain in normal brain tissue under most conditions. It appears that the individual neuronal generators lack synchrony except in epileptic seizures and other disordered states of nervous tissue.

At this point, I shall present some numbers that relate to what Dr. Kalmijn said earlier about the sensitivity of the electroreceptor mechanisms in the fish (see Chapter 2). The figure that he gave was 0.01 μV/cm for the field that could be sensed by the sharks and the rays. In brain tissue the EEG is on the order of 50 μV/cm, large by comparison with the above field but small by comparison with the 1 kV/cm transmembrane change in potential that characterizes a synaptic impulse. The membrane potential itself is on the order of 100 kV/cm. This potential is close to dielectric breakdown, and obviously it would be very difficult to make an artificial dielectric that would have that strength without risk. So the question of whether the EEG surrounding the brain cell is sensed by the cell and thus has some informational content is related to the question of whether or not imposed environmental fields are likely to cause a change in brain function.

More or less as a parenthesis but to express a point of view, I want to show that a constancy of EEG pattern can be detected in specific behavioral performances. Because the EEG is a complex record, people have been drawn to examine it visually and express opinions about it. Over the years, pattern recognition methods have been developed which show that EEG patterns indeed have the nature of signatures if one looks at them appropriately. For example, one can ask human subjects questions that have psychologically high-stress content (e.g., relating to sexual behavior, truthfulness, pornographic reading material, etc.), and then ask the same subjects questions that would be regarded as low-stress.[3] Examples of such questions are shown in Table 1. If one carries out this procedure in a well-defined manner — first asking a question, recording the time when the subject answers, asking another question, recording the response time, and so on — one finds that the EEG does show brief alpha-like bursts across many of the scalp leads immediately following the subject's response, as demonstrated in Figure 3. Obviously this is not something that allows the recognition of a pattern. If, however, one does spectral analyses and then uses a discriminant-analysis method of looking at the intrinsic patterns, one finds that for each individual it is possible, with a very high order of accuracy (in this case over 90%), to classify those EEG records into the types of questions that were being asked. This is not unique to an individual. In this particular study, we used many subjects, recognizing their individual signatures in the EEG and also the signatures of groups of individuals. Thus, something very interesting is happening in the EEG which we can recognize with appropriate techniques. This leads to the question of whether a change in behavior of some subtle kind could be seen if one were to impose environmental fields on the head that looked like the EEG or were in the ELF range of EEG frequencies. We

Table 1. Description of question sets. Autonomic criteria are based on evoked circulatory responses to each item, averaged across an independent group of sixty-five subjects.

Question Set	Autonomic Criteria	Typical Verbal Content
High Stress I (Sex)	Severe (±2 std dev) fluctuations of heart rate and pulse volume coincident with question presentation.	"Do you masturbate?" "Do you have homosexual urges?" (4 items)
High Stress II (Lie)	As above.	"Have you lied answering those questions?" (repeated 3 times)
Low Stress I (Read)	Mild (less than 1 std dev) fluctuations of heart rate and pulse volume coincident with question presentation.	"Do you approve of the material you have just read?" (3 items)
Low Stress II (Query)	As above.	"Does anyone know you are here?" (repeated 2 times)
Non-Stress	No fluctuations of heart rate or pulse volume noted.	"Do you believe in capital punishment?" (2 items)

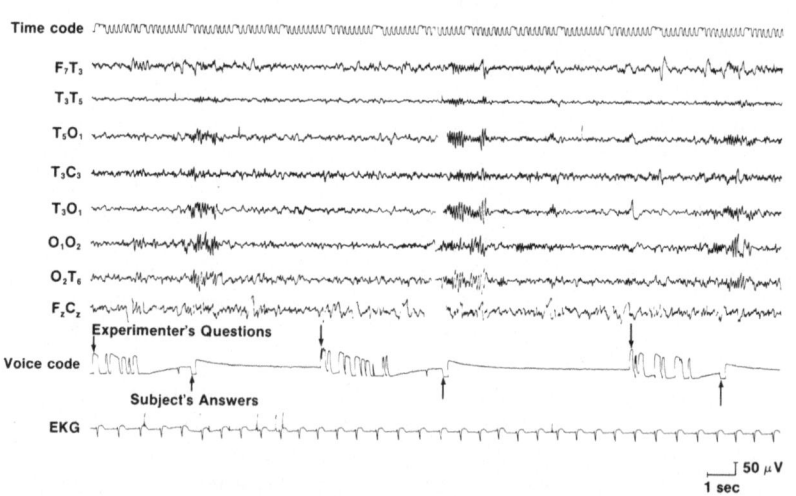

Figure 3. EEG records from a subject during questioning from a taped protocol. Note alpha-like burst of waves following response to each question. (From Berkhout, Walter and Adey, 1969. *Electroencephalogr. Clin. Neurophysiol.* 27:457-469; reproduced by courtesy of Elsevier/North-Holland Biomedical Press.)

know from Dr. Kalmijn's work that the marine vertebrates have this strange sensitivity involving their special sense organs, the ampullae of Lorenzini.

About 15 years ago, groups in various parts of the world became interested in the possibility that exposing man or higher animals to environmental fields might cause subtle changes in behavior. Some work done in Germany by R. Wever involved studies of circadian rhythms in human subjects housed in bunkers that were screened electrically against environmental fields.[4] Wever observed the daily rhythms of a subject over a period of 20 days. The daily rhythm of sleep and wakefulness was plotted in the absence of the field, and Wever observed under these conditions that the 24-hour rhythm had become free-running, with a period of about 26.6 hours. On the seventh day, a 10-Hz square wave electric field with an amplitude of 2.5 V/m was imposed on the subject without his knowledge, and the daily rhythm moved back to 24.8 hours. On the 16th day the field was withdrawn and the subject went off into a curious rhythm which for sleep and wakefulness was 36.7 hours, whereas the temperature and heart rate measurements indicated a retention of something approaching a 24-hour rhythm. This was apparently a case of internal dissociation.

Wever's work, which immediately became controversial, led to the interpretation that the presence of the normal components of low frequency fields in the environment is necessary for the retention of circadian rhythms. The level of these environmental fields varies with stages of the sunspot cycle, an 11-year cycle; at the height of the cycle it approaches 1 mV/m at 10 Hz, indeed a very low figure. Wever went on to include birds in his study, since these animals are less likely than humans to have motivations that could bias the experimental results. He obtained the interesting result that birds showed the same phenomenon. A 24-hour periodicity was achieved in the circadian rhythm under the influence of the field. This further confirmed that indeed the 10-Hz, 2.5-V/m square wave field used by Wever has some driving capacity.

It was at about the same time in the middle 1960's that our own laboratory became interested in this problem. James Hamer, a physicist from Northrop, informed us that he had learned from the Eastern European literature that in the years of highest sunspot activity there was an increased number of deaths from crimes of violence and stress-related diseases recorded in the public health actuarial statistics. At first I was completely disbelieving, but after a year and a half, Hamer was able to persuade us to do an experiment recording simple reaction times in subjects exposed in a double blind experiment to fields between 5 and 15 Hz at 2 to 10 V/m. Hamer did see a statistically significant shortening in the reaction time, by about 5 msec, in the 200-msec average reaction time recorded for 59 subjects.[5] But the data were not clear, and some of us felt that if he were correct, a more appropriate experiment should be done.

This led to the next study, continued to this time under Dr. Rochelle Gavalas-Medici, who has tested a monkey's ability to estimate the passage of a 5-second time interval without cues. It takes about four months to train a monkey to do this, using apple-juice to reward the animal if its response is within ±1 second of the 5-second interval. The experiments were done in animals with and without implanted electrodes.[6]

The Navy performed a fairly accurate measurement of the total current dumped in the head of a phantom monkey by using a special sensor with a light pipe connection to the external measuring apparatus. The current present in the head of this phantom was 0.9 nanoamps. If, for the sake of making an order-of-magnitude calculation, one assumes a uniform distribution of the current through the head, then one arrives at a gradient of 10^{-7} V/cm at 7 Hz for a 10-V/m peak-to-peak field. This is almost down to the level that Dr. Kalmijn was talking about in marine vertebrates. One sees in most circumstances that these fields will produce a shortening in the subjective estimate of time passage (Figure 4). Briefly, three field exposure situations are shown in the figure: 1 V/m, 10 V/m, and 56 V/m. All these fields are oscillating at 7 Hz. For the interresponse times shown, at 1 V/m the graphs for the control distribution and the test distribution are almost superimposed. At 10 V/m the field is starting to produce a shift to the left in the solid line, the "field-on" situation. At 56 V/m there is a clear shift of about a half second in the estimate of a 5-second interval. This field produces approximately 10^{-7} V/m in the brain. Because that is a very low-level field,

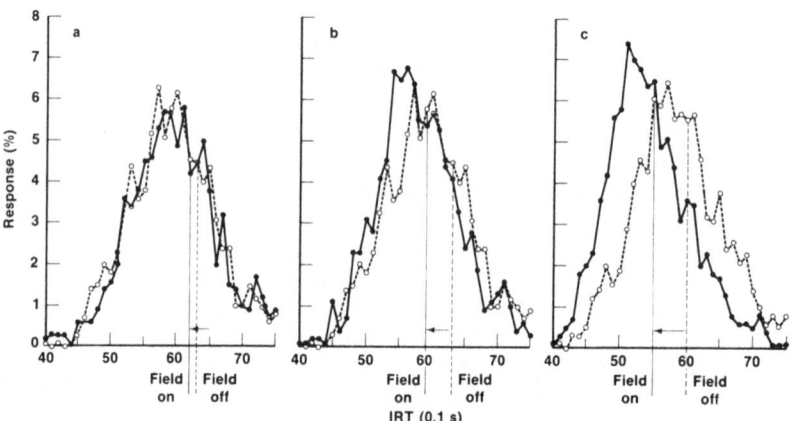

Figure 4. Distribution of interresponse times (IRTs) in a 4-hr experiment with one monkey during exposure to a 7-Hz sinusoidal field at (a) 1 V/m, (b) 10 V/m, (c) 56 V/m. The monkey was trained to estimate a 5.0- sec time interval without sensory cues. Field and no-field distributions are contrasted (solid lines, field exposure; dashed lines, control tests). Numbers on abscissae indicate 100- msec "bins" used in computations (for example, number 60 indicates an interresponse time of 6.0 sec). This animal had no implanted EEG electrodes, nor any other metal in its cranium. There was a progressive shortening in interresponse times (shown as a shift to the left) as the field strength was increased to 10V/m (b) and 56 V/m (c). (From Gavalas-Medici and Day-Magdaleno, 1976, *Nature (London)* **261**:256-259; reproduced by courtesy of Macmillan Journals Ltd.)

we wondered whether clearer effects might be obtained with fields closer to the EEG gradient of about 50 mV/cm. It is possible to induce a field of that amplitude by using radiofrequency carriers and amplitude-modulating them at EEG rates; therefore, we have carried out many studies with radio-frequency signals at 147 MHz and 450 MHz on which an amplitude-envelope-modulation is imposed at frequencies between about 1 and 35 Hz.

Some of the first experiments were done by Dr. Bawin[7] on the EEG rhythm signatures of the cat. The awake cat shows rhythm signatures every few seconds in deep brain structures and in some cortical regions. It has long been known that it is possible to condition the appearance of these signatures. In other words, an animal shown a flash of light will show the rhythm sig-nature in the EEG within about two seconds. If the rhythm fails to appear, the cat is given an aversive or punishing stimulus, either inside the brain or else-where in the body. In these experiments the conditioning was carried out in such a way that if the animal, after seeing a flash of light, did not make the EEG rhythm that we desired, its eyes were caused to deviate involuntarily to the opposite side by stimulating the frontal eye field. Cats really dislike this procedure, and within about a day or so they learn to make the rhythm to avoid the unpleasant (but not painful) experience of having the eyes involuntarily deviated.

Figure 5 shows the responses to training in a radiofrequency field. By the fifth conditioning session the animal's performance level is at 63% and there is some evidence of rhythms starting to build up following the flash of light. By the tenth conditioning session, when the animal sees the flash it is strongly motivated to produce these rhythms within about a second after the flash. One can observe the rate at which the animal learns this, and one can also extinguish the response in the fashion of behaviorists, namely, if the animal is not punished, the response will tend to disappear. The field is given the same modulation rate as the particular signature that we wish to reinforce. The signatures have many different frequencies, varying from 2 to 14 per second, but the radiofrequency signal can be amplitude-modulated at the same frequency as the rhythm signature which we have selected for examination.

Some of the typical results from testing in a radiofrequency field will now be shown. Figure 6 shows results for animals trained and extinguished with and without the presence of the radiosignal. The cat trained in the absence of the field comes up to a stable level of about 70% correct responses and then, when it is no longer punished for not making the response, drops back to the pre-training level within a day or so. We find that the animal trained in the presence of a radiosignal attains a much higher level of performance in the presence of that radiosignal, up to something like 95%. In the extinction trials where the animal is no longer punished if it does not make the response, it goes on making the correct response for almost two months, or about 46 training sessions after the punishment has been withdrawn. In other words, there appears to be a persistent memory in the presence of the radiosignal.

Very interestingly, as a rather important technical parenthesis, the modulation signal on the EEG leads appears only when the animal makes its

Operant conditioning of specific
Electroencephalographic patterns - Cat No. 5

Preconditioning session

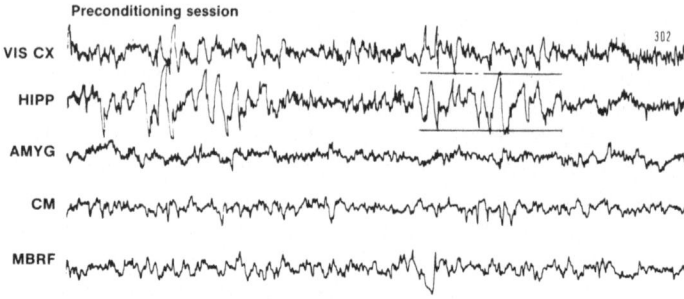

VIS CX

HIPP

AMYG

CM

MBRF

5th conditioning session - level of performance 63%

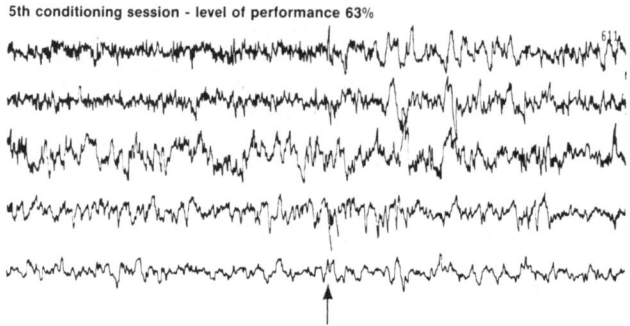

10th conditioning session - level of performance 85%

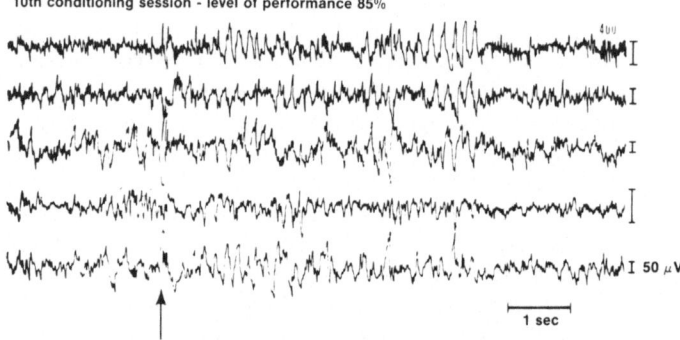

1 sec

Figure 5. Conditioning of EEG rhythmic activity to a light flash (arrow) was initially seen in visual cortex (VIS. CX) and hippocampus (HIPP). Failure to produce this response within 2 sec was followed by an aversive or punishing brain stimulus that caused involuntary turning of the eyes (see text). (From Bawin, S.M., 1972, Cat EEG and behavior in VHF electric fields amplitude-modulated at brain wave frequencies. Ph.D. thesis, Department of Anatomy, University of California at Los Angeles.)

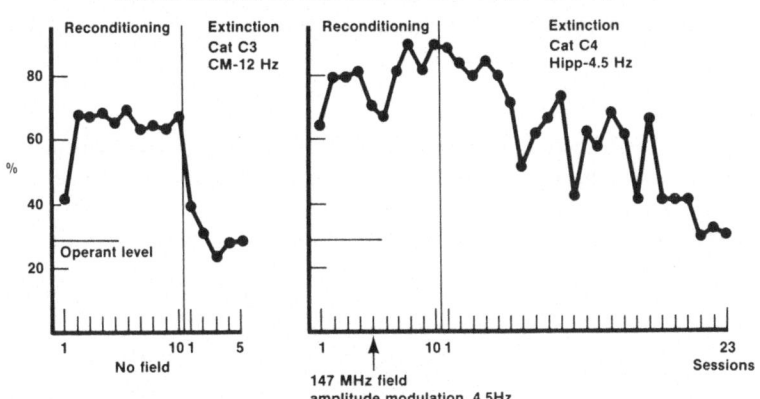

Figure 6. Training and extinction trials without (left) and with (right) 147-MHz, 0.8-mW/cm^2 field, amplitude-modulated at frequency of an EEG rhythm burst or "signature" in that particular brain structure. In the extinction trials, no aversive stimuli followed incorrect responses to light flash. Presence of the modulated RF field greatly prolonged the extinction of the learned response. (From Bawin, Gavalas and Adey, 1973, *Brain Res.* **58**:365-384; reproduced by courtesy of Elsevier/North-Holland Biomedical Press.)

own response. We are not overwhelming the cat's brain through the electrode system with a large artifactual rectification of the applied radiofield. Tests by spectral analysis make it obvious that the cat only produces the rhythm, or it only appears in the spectrum, when there is an actual physiological response. That leads to the question of the domain size in which signal detection is occurring; it must be very much smaller than the distance between the electrodes, which in this case was on the order of about 0.5 - 1 mm. This, then, is evidence that one can modify behavior by the presence of radiosignals made to look in some degree like the brain's own signaling system.

Soviet research in this area had gone a long way before workers in the United States became interested in it. Recently, after two years of negotiation, I received a Russian instrument called the LIDA, model 4. This device was designed by Dr. Rabichev in Soviet Armenia for "the relief of psychoneurotic illness" and it puts out about 40 to 80 W of radiation at 40 MHz, pulsed at rates up to 100 per minute. These are long pulses lasting 100 msec. Concomitantly one may, if one wishes, use pulsed light and pulsed sound, though the efficacy of the apparatus does not appear to be related to the additional sensory units. (In the course of the correspondence required to secure this instrument, I discovered that it is covered by a very large U.S. patent taken out by the Russians.) We have not tested the instrument yet, but some of the photographs given to us show a hall full of people, all of whom fell sound asleep within a matter of 15 minutes after the instrument was turned on. While I was in the Soviet Union a year or so ago, some of the newspaper correspondents I talked with indicated that the use of this type of device in a classroom situation is being considered. It was suggested in these discussions that it might be feasible to use this device with mentally retarded children in the hope of enhancing attention.

The question of how these fields may be sensed is the next subject that I wish to pursue. First of all, the brain must be considered as a tissue with three compartments.[8] There is a neuronal compartment, a surrounding neuroglial compartment composed of cells which outnumber the nervous element by a factor of five, and between the two is a very curious but well-organized material that is protein in nature (Figure 7). This material appears to be

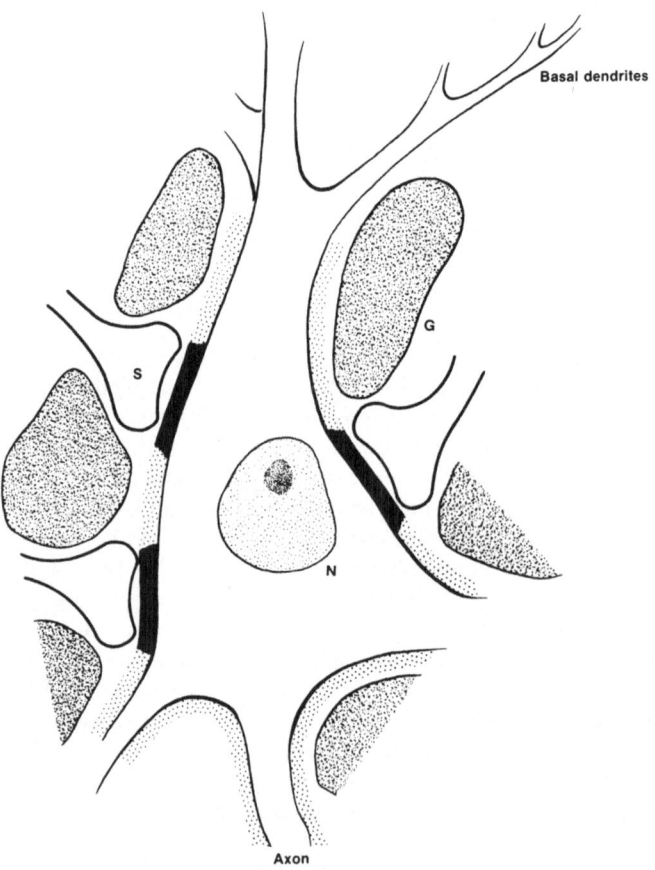

Figure 7: Model of brain tissue, with two cellular compartments: neuronal and neuroglial. A third compartment, the extracellular space, lies between the two cellular compartments, and contains stranded macromolecular terminals of proteins that lie as intramembranous particles (IMPs) within the lipid bilayers that constitute the classic cell membrane. These cell surface protusions form polyanionic, glycoprotein sheets that act as receptor sites in the transductive coupling of weak electrochemical stimuli (see text). G = glial cell; S = synapse; N = nucleus of neurona.

especially densely organized in the vicinity of synapses, and its appearance is best demonstrated in brain preparations stained with phosphotungstic acid at low pH (around 1.25). One might expect that the large tetravalent phosphotungstic acid molecule would not compete successfully with high concentrations of hydrogen ions for binding sites at the cell surface, but that is not the case. The surface plaques of the material appear incomplete, but in general cover much of the cell surface, particularly on dendrites.

One can consider the cell membrane as having a lipid bilayer enclosed in a sandwich of structural protein, with external protrusions of glycoprotein whose strands terminate in negatively charged amino sugars.[9,10] These strands are the protrusions of intramembranous particles (IMPs). The protrusions and the lipid bilayer make up the "greater membrane," because its depth may be hundreds of angstroms, whereas the lipid bilayer alone may be about 40 Å across. In this model, the IMP has an external protrusion and also an internal protrusion that makes contact with microfilaments in the cell. These microfilaments in turn may make further connections with microtubules that lead directly to the nucleus. With such a model, if it is correct, one would have a direct communication system for chemical or electrochemical sensing from outside the cell more or less directly to the nucleus.

It has also been demonstrated that the IMPs are movable in the fluid mosaic of the lipid bilayer. This mobility suggests a scheme of longitudinal organization in the membrane, which is an important point that should be stressed. In our concepts of transductive coupling, we now think more often of events as occurring along the membrane in length and in area rather than through the membrane in the transverse sense as most excitatory processes have traditionally been viewed. If this surface is polyanionic, it will tend to bind cations and, as Katchalsky first pointed out in the middle 1960's and others have demonstrated subsequently, there are two ions in particular that will bind to the external strand − calcium and hydrogen ions. Both are relevant to the line of thought that I want to develop in the remaining portion of this discussion.

Calcium can be seen in light microscopy sections stained by a technique that uses a tetravalent acetate chelating agent. The neuron body is seen as a blank space surrounded by pinkish dots. In dendrites the pinkish dots extend much like synaptic boutons along the dendritic shaft. The membrane itself is also faintly outlined by the pink stain, indicating the presence of calcium. By contrast, much of the material around the neuroglial cell and its cytoplasm is a uniform pink, suggesting that there is a lot of calcium in it as compared to the neuron. So we have a distinctly different relationship in the distribution of calcium on the surface of the nerve cell, probably associated with synaptic organization.

This was a thesis study by Theodore Tarby in my laboratory in 1968, and we pressed on at a later time to see if we could move some of this calcium by the application of an electric field.[11] In studies with Dr. Kaczmarek, we stimulated the intact awake cat cortex through large agar electrodes which separated the metal-tissue interface by several centimeters, thus insuring a very uniform electric gradient. We used pulse trains at around 200 per second for most of these experiments, but the actual frequency chosen did not appear to matter over the range from about 10 to 200 per second. The electric gradients that we applied were on the order of 50 mV/cm, the same as the EEG. As a result of electric stimulation we released a very large amount of calcium leading to about a 20% increase in the usual efflux, and we also released GABA, the amino acid transmitter substance, in almost equivalent amounts.[12] This finding exemplifies the fact that a gradient as weak as the EEG is associated with a major change in the efflux of an ion essential to the excitatory process and to the release of a transmitter substance.

We pressed on from this type of study to see if externally-imposed fields alone could alter the efflux of calcium. In a study with Dr. Bawin we used ELF fields of the same strength as those that had produced behavioral changes in monkeys, namely, in the range from 5 V/m to 100 V/m in air.[13] The frequencies of these fields are shown in Figure 8, and range from 1 to 32 Hz. Very briefly, there was a major decrease in the efflux of calcium by about 15 to 20% at frequencies of 6 and 16 Hz for fields of 10 and 56 V/m, but relatively little at 5 V/m and 100 V/m. At this point we became aware that the process did not have the classic aspects of an equilibrium interaction. Something strange was happening in which the efflux of calcium was constrained in both the frequency and amplitude domains so that we started to observe windows. These windows also became apparent when we used stronger radiofrequency signals that produced fields of 10^{-3} to 10^{-2} V/cm in tissue, which is comparable to the 50 mV/cm EEG signal.

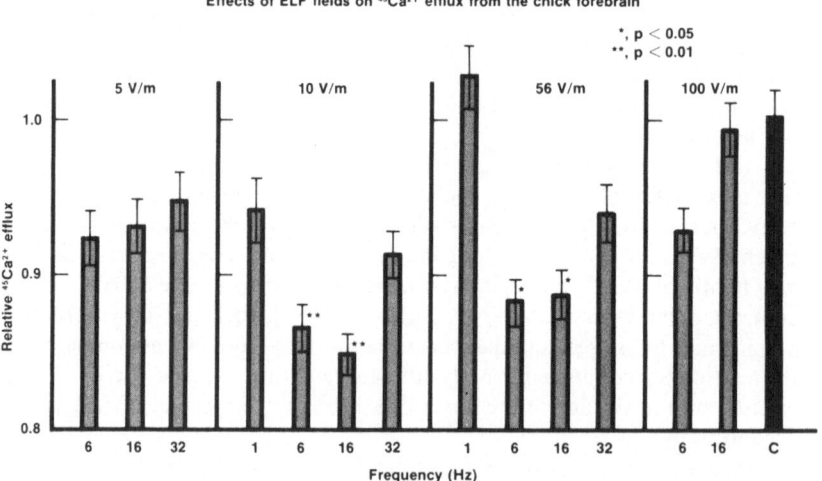

Figure 8. Effects of extremely low frequency fields on $^{45}Ca^{2+}$ efflux from chick forebrain. The relative $^{45}Ca^{2+}$ effluxes are referred to the mean control value (C). Error bars represent ± 1 standard error of the mean. *, $p < 0.05$; **, $p < 0.01$ that the mean is the same as that of the control. (From Bawin and Adey, 1976, *Proc. Natl. Acad. Sci. USA* 73:1999-2003.)

We have carried out these electromagnetic field experiments in a variety of ways. We have a horn radiator that can produce from 0 to 20 mW/cm² at 450 MHz frequency. The amplitude of this field is sinusoidally modulated at 1 to 30 Hz. The chamber contains an orthotropic volume at the far end of the horn of about 2 m³ where one can place cats or monkeys or tissue. We can heat the chamber to 37°C for studies with isolated tissues, thus providing a very desirable way to test either the intact animal or the isolated tissue. The standing wave ratio is less than about 1.2 to 1 over this 2 m³ volume at the far end of the horn.

Figure 9 shows one of the first studies that we did with radiofrequency signals having the characteristics of a biological low frequency signal because of imposed amplitude modulation. One sees now an *increase* in efflux of calcium as a function of modulation frequency, not the *decrease* we saw with the ELF fields.[14] This was an increase up to 20% using the no-field efflux as a reference; unmodulated, the radio signal is without effect, but as the modulation frequency increases, through the range from 6 to 20 Hz, there is a sharp and very significant increase in the calcium efflux which then declines and disappears at 35 Hz. In this study, which was initiated by Dr. Bawin, the windowing in the frequency domain has its counterpart in the power domain because these effluxes only increase at incident power densities between 0.1 and 1 mW/cm² (Figure 10). Both increases are significant at the $p < 0.05$ level. Below 0.1 mW/cm² and above 1 mW/cm², the efflux either is negative or there is an insignificant increase.[15]

About three years ago, the Environmental Protection Agency became very interested in these studies and set up their own laboratory essentially to do counterpart but quite independent experiments conducted under the

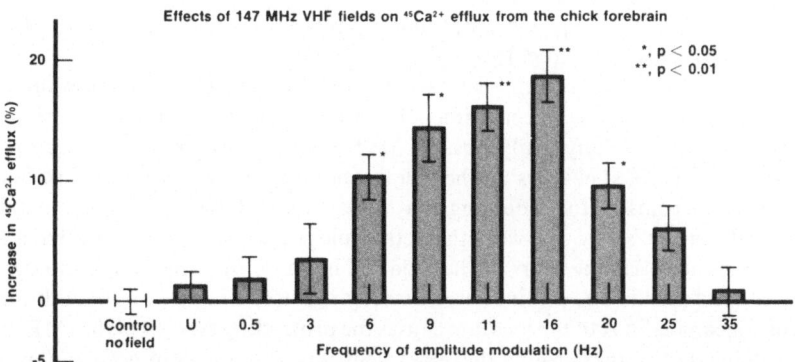

Figure 9. Effects of amplitude-modulated 147-MHz VHF fields on the $^{45}Ca^{2+}$ efflux from the isolated neonatal chick forebrain. The efflux is expressed as the percent increase relative to the control condition (without an applied field). Error bars represent ± 1 standard error of the mean. An unmodulated field (U) had no significant effect on calcium efflux relative to the no-field control. *, $p < 0.05$; **, $p < 0.01$. (From Bawin, Kaczmark and Adey, 1975, *Ann. New York Acad. Sci.* **247**:74-80; reproduced by courtesy of the New York Academy of Sciences.)

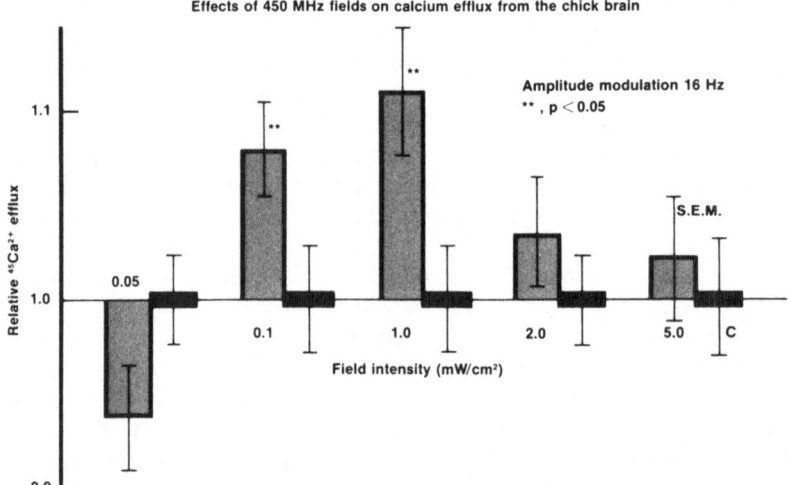

Figure 10. Effects of changing incident power of a 450-MHz field, amplitude-modulated at 16 Hz. Significantly increased $^{45}Ca^{2+}$ efflux from isolated neonatal chick forebrain only occurred for incident field intensities in the range from 0.1 to 1.0 mW/cm^2. (From Bawin, Sheppard and Adey, 1978, *Bioelectrochem. Bioenergetics* 5:67-76; reproduced by courtesy of Birkhauser Verlag.)

direction of Dr. Blackman and Dr. Elder.[16] They have confirmed all aspects of the findings, including both the power windowing and the frequency windowing. A sharp intensity window was observed in their experiments, with increased calcium efflux occurring only around 0.75 mW/cm^2, with insignificant effects above and below this intensity. They have also seen sensitivity to low-frequency amplitude modulation, with the maximum effect occurring between 9 and 16 Hz.[16]

Given that there are these windows, why are they there? I want to conclude with some speculations that might initiate critical discussions. First of all, it is physically possible that there are long-range interactions between the fixed charges on the membrane surface glycoproteins. The first events in transductive coupling may thus occur along the surface of the membrane. It is here that work in electrobiology parallels much work that has been done in recent years in the fields of immunology and endocrinology. For example, as Dr. Malinin mentioned earlier (see Chapter 5), the addition of concanavalin A to the medium causes the protruding strands of the IMPs to congregate together around the tetravalent concanavalin A molecule. This is the counterpart of what occurs when the antibody molecule binds on the surface of lymphocytes and the IMP strands are caused to move into patches and later into a cap.[17] The process occurs along the length of the membrane through the building of blocks of protein around calcium and magnesium ions, so that a bridge is built from the Y-shaped antibody molecule back to the surface of the cell. This process is cooperative, that is to say, a weak event at one point causes a series of changes to be triggered along the length of the membrane. It is a nonequilibrium process, occurring in systems which, in a sense, have already spent energy to prepare for this type of event. The evidence for cooperativity reported by Yahara and Edelman[17] was related to a sharp temperature coefficient for the capping and patching phenomenon that I just described.

Immunologists some years ago had moved into this area of cooperative processes which, by definition, state that a very weak trigger at one point can be the trigger for some remote event. Similarly, endocrinologists have been aware that the binding of a hormone molecule at the surface of a cell within the thicket of glycoprotein strands produces changes that are widespread, and perhaps may cover the whole cell surface. These ripples can extend far and wide. It has been shown that as few as two molecules of prostaglandin E_1 on the surface of a red cell can affect by as much as 30 to 40% the rate at which these cells will drip through small pores in a filter.[18] In other words, there is a signaling over a very wide area of membrane upon arrival of a molecule of this kind.

This leads us down to a finer level of organization in our studies of electromagnetic field interactions. It might be supposed that there are coherent patches established on the surface of the membrane in the presence of the field. Gerhard Schwarz first demonstrated in 1967 that on a biopolymer sheet, there will be areas in which the adjacent charges all have equivalent energies and in that state will be subject to less thermal perturbation than the surroundings.[19] This area then could be a preferred site for the binding of calcium ions. In order for this to happen, energy must be provided, and as shown in the studies of Yahara and Edelman, much of that energy is metabolically expended in achieving the receptor site energy states for the binding event. One can also suppose that in the binding of cations like calcium, a certain energy is expended from within the cell to achieve a coherent patch on the membrane surface. And so, calcium could become bound to the sites that are coherent, as Schwarz observed, for periods running out into the millisecond range. Schwarz's test system involved poly-1-glutamic acid and the binding of acridine dyes. Additionally, electric fields of the type used in our experiments may indeed move calcium from these binding sites in a cooperative process, so that one would see the cooperative activity spreading beyond a single patch and involving wide areas of the membrane surface.

I shall now briefly discuss models that have appeared during the last few years bearing on this question of cooperativity at membrane surfaces in the sensing of weak fields. A model proposed by Grodsky[20] envisages the membranes as having lines of electric strain coming out from the lipid bilayer and the glycoproteins having dipoles arranged essentially in an antiferromagnetic state in the field. Grodsky's model deals in a quantum-mechanical sense with phase transitions, of the zero order Einstein-Bose type, occurring as energy is added to that system by a low frequency disturbance. I do not presume to be in any way cognizant of the details of quantum mechanics, but the phase transition diagram that Grodsky addresses is one in which the triple point (the null point) can be considered as separating these dipoles into three basic states: the antiferromagnetic state, the flop phase, and the paramagnetic state. Energy added in a coherent fashion will cause the dipoles scattered about in the surface glycoprotein to change their orientation from the anti-state to the flop-phase, and finally at much higher energy levels take on a para state. Grodsky's model leads to the conclusion that, at the zero order phase transition point, the sysem could take up energy in a resonant fashion at very low frequencies on the order of biological frequencies. A huge amount of energy could be dumped out of the system as this resonance grew in amplitude. The weakness of the model, as I understand from listening to discussions of it, is that the temperature range over which this would occur would be extremely narrow indeed, on the order of 1/10 to 1/100 of a degree.

Other models, proposed by Fröhlich[21] and by Kaczmarek,[22] speak to the possibility that energy may be taken up by proteins when a very high frequency field is imposed. Fröhlich suggests that exposure to fields in the range of 10^{11} - 10^{12} Hz would cause some of these protein molecules to move above the ground energy state. The absorption of this energy would then depend upon a series of long-range interactions. The most recent development in his thinking gives a Volterra equation approach to ways in which populations of activated and ground state proteins would be exchanging in the system.

Kaczmarek has considered the same model with a limit cycle type of shift, namely, the protein molecules move from the ground state up around the loop of the limit cycle and back to the ground state again. At a more earthy level, many of the questions about interactions of this type have been addressed to the question of whether or not the field being sensed in the tissue is at or below the background of the thermal noise level. Obviously, if the interaction is occurring below the thermal noise level, some quite remarkable levels of cooperativity would be necessary.

Thermal noise at physiological temperatures is about 0.02 eV. That fact says little about what is significant to the brain in the transmission of information from one part of the neuron to another. It does say that if we had infinite noise bandwidth in the system, a molecule would need to be about 300 meters long before a field of 10^{-8} V/cm could be detected. Obviously, we don't have molecules 300 meters long. However, the surface of the membrane may be acting as a low pass filter for the transfer of thermal noise. (This is a model that is much more familiar to the engineer than it is to the physicist apparently, and one can express the noise as a transfer function relating to the bandwidth of the system in which it is being transferred, and the specific resistance.)

My colleagues and I have had some interest in this type of model because if one looks at the surface of the membrane in the counter-ion layer, that is to say within 20-50 Å of the surface of the membrane, it may indeed be functioning as a low pass filter. Einolf and Carstensen[23] studied the behavior of micron-sized particles that have a porous surface through which lines of electric strain protrude. Theirs was an extension of the Schwarz model of the membrane surface to the porous condition. First in their model, and then in actual tests, they saw that there is a dielectric dispersion with an apparent dielectric constant in excess of 10^6 at frequencies of less than 1.0 kHz in the counter-ion layer. It is indeed almost an ideal low pass filter. Taking that figure and a specific resistance of 300 ohms/cm for the brain, one can calculate that to displace ions along the surface, the transfer function for the bulk of thermal noise would be 10^{-8} V/cm,[12] which is exactly the figure that Dr. Kalmijn has indicated as the biological sensitivity in the marine vertebrates. This is also within an order of magnitude of the sensitivity observed for the mammalian central nervous system in the ELF experiments, both chemically and behaviorally.

But there are other aspects to the problem that are even more interesting, such as the question of an amplitude window. There is the possibility that some form of tunneling is occurring along the surface of the membrane. This tunneling may be occurring between the charges in the coherent and noncoherent regions, and across the energy barrier or phase partition. A number of scientists have been interested in the possibility of biological tunneling. John Hopfield[24] has a rather beautiful model of the hemoglobin molecule based on electron tunneling. What I want to suggest is that we may be seeing here not electron tunneling but proton tunneling. The requirements for charge density and the tunneling distances are well within the physical confines of what we know about the surface of the membrane. Additionally, as was suggested ten years ago by Bass and Moore[25] working in Pauling's laboratory at Cal Tech, perhaps the first change in excitation is the displacement of calcium ions by hydrogen ions to immediately adjacent binding sites. This leads to a depletion of the hydrogen ion concentration in the local atmosphere and is expressed as a local alkalosis. We have evidence from our most recent experiments, not yet published, that this may indeed be the case because it appears that the field sensitivity is very much a function of the hydrogen ion concentration.

So I leave this rather speculative discussion with you, pointing out that there are apparently special structural organizations in the central nervous system of all vertebrates and particularly of the mammal, which may make him susceptible in very subtle ways to the environmental electric fields, electromagnetic fields, and to the intrinsic fields in brain tissue. I certainly do not believe that we can dismiss the EEG as the noise of the brain's motor.

REFERENCES

1. Schmitt, F. O., Dev, P., and Smith, B. H., 1976. Electrotonic processing of information by brain cells. *Science* **193**:114-120.
2. Elul, R., 1972. The genesis of the EEG. *Int. Rev. Neurobiol.* **15**:227-272.
3. Berkhout, J., Walter, D. O., and Adey, W. R., 1969. Alterations of the human electroencephalogram induced by stressful verbal activity. *Electroencephalogr. Clin. Neurophysiol.* **27**:457-469.

4. Wever, R., 1968. Einfluss schwacher electromagnetischer felder auf die circadiane periodik des menschen. *Naturwissenschaften* **1**:29-33.

5. Hamer, J., 1969. Effects of low level, low frequency electric fields on human reaction time. *Commun. Behav. Biol.* **2**(A):217-222.

6. Gavalas, R. J., Walter, D. O., Hamer, J., and Adey, W. R., 1970. Effect of low-level, low frequency electric fields on EEG and behavior in *Macaca nemestrina. Brain Res.* **18**:491-501.

7. Bawin, S. M., Gavalas, R. J., and Adey, W. R., 1973. Effects of modulated VHF fields on the central nervous system. *Brain Res.* **58**:365-384.

8. Van Harreveld, A., Crowell, J., and Malhotra, S. K., 1965. Extracellular space in the cerebral cortex of the mouse. *J. Cell Biol.* **25**:117-137.

9. Adey, W. R., 1977. Models of membranes of cerebral cells as substrates for information storage. *BioSystems* **8**:163-178.

10. Singer, S. J., and Nicholson, G. L., 1972. The fluid mosaic model of the structure of cell membranes. *Science* **175**:720-736.

11. Adey, W. R., 1970. Cerebral structure and information storage. *Prog. Physiol. Psychol.* **3**:138-201.

12. Kaczmarek, L. K., and Adey, W. R., 1973. The efflux of $^{45}Ca^{2+}$ and [^{3}H] gamma-aminobutyric acid from cat cerebral cortex. *Brain Res.* **63**:331-343.

13. Bawin, S. M., and Adey, W. R., 1976. Sensitivity of calcium binding in cerebral tissue to weak environmental electric fields oscillating at low frequency. *Proc. Natl. Acad. Sci. USA* **73**:1999-2003.

14. Bawin, S. M., Kaczmarek, L. K., and Adey, W. R., 1975. Effects of modulated VHF field on the central nervous system. *Ann. N. Y. Acad. Sci.* **247**:74-81.

15. Bawin, S. M., Sheppard, A. R., and Adey, W. R., 1978. Possible mechanisms of weak electromagnetic field coupling in brain tissue. *Bioelectrochem. Bioenergetics* **5**:67-76.

16. Blackman, C. S., Elder, J. A., Bennane, S. G., Weil, C. M., and Eichinger, D. C., 1977. Two factors affecting the radiation-induced calcium efflux from brain tissue, *Symposium on the Biological Effects of Electromagnetic Waves, Airlie, Va., Oct. 30 - Nov. 4.*

17. Yahara, I., and Edelman, G. M., 1972. Restriction of the mobility of lymphocyte immunoglobulin receptors by concanavalin A. *Proc. Natl. Acad. Sci. USA* **69**:608-612.

18. Allen, J. E. and Rasmussen, H., 1971. Human red blood cells: Prostaglandin E₂, epinephrine, and isoproterenol alter deformability. *Science* **174**:512-514.

19. Schwarz, G., 1967. A basic approach to a general theory for cooperative intramolecular conformation changes of linear biopolymers. *Biopolymers* **5**:321-324.

20. Grodsky, I. I., 1976. Neuronal membrane: A physical synthesis. *Math. Biosci.* **28**:191-220.

21. Fröhlich, H., 1968. Long-range coherence and energy storage in biological systems. *Int. J. Quantum Chem.* **11**:641-649.

22. Kaczmarek, L. K., 1977. Cation binding models for the interaction of membranes with EM fields. *MIT Neurosci. Res. Prog. Bull.* **15**:54-60.

23. Einolf, C. W., and Carstensen, E. L., 1971. Low frequency dielectric dispersion in suspensions of ion-exchange resins. *J. Phys. Chem.* **75**:1091-1099.
24. Hopfield, J. J., 1973. Relation between structure, cooperativity and spectra in a model of hemoglobin action. *J. Mol. Biol.* **77**:207-222.
25. Bass, L., and Moore, W. J., 1968. A model of nervous excitation based on the Wien dissociation effect. In *Structural Chemistry and Molecular Biology*, eds., A. Rich and C. M. Davidson, pp. 356-368. San Francisco: W. H. Freeman and Company.

DISCUSSION

Q: In one of your remarks, you were discussing experiments in which the brain of a monkey was stimulated with an electric field. Have you measured the current density and field strength using a phantom?

A: We have obtained an estimate of what the current density might be in the head as a result of the field capacity being coupled from the large plates on the side of the head. This measurement indicated total current to ground in a phantom of the same size as a real monkey's head and with similar electrical properties. Current density was 0.9 nanoamps for an applied field of 10 V/m in air. Therefore, if one takes about 300 ohms as the specific resistance of brain tissues, the calculated voltage gradient is 10^{-7} V/cm. It is a simple capacity model of the coupling between the head and the plates.

Q: Would you care to speculate in more definite biochemical terms as to what regions of the membrane may act as a low pass filter? Are you speaking in terms of glycoprotein or glycolipid components?

A: I am speaking generally about the material that has the highest density of anionic charges and these are the terminal sialic acid residues at the ends of protruding strands of IMPs. The strands themselves are structurally of many varieties because they range in molecular weight from a few hundred to about 50,000 daltons, but they form a huge forest of projecting strands, most of which are negatively charged in the normal cell membrane. One can consider the sheet arrangement of the membrane as broken up by the protrusions of material from inside the membrane. Gerhard Schwarz thinks that the sheetlike behavior of the glycoproteins may be significantly interrupted by the presence of glycolipids which have a very much lower charge density. In general, however, the protruding strands are either glycoproteins or glycosoaminoglycans.

Q: In more specific terms, what I am asking is how the electromagnetic signal is transmitted through the membrane?

A: It is possible that the signal is transmitted through an induced allosteric change, or twisting, of IMP molecules. In that twisting, calcium is a required ion because its shifting from one binding location to another is what causes the molecule to either coil or uncoil. Many macromolecules can change volume by as much as 100,000 times without alteration in molecular structure merely by the addition or removal of calcium. You can play this game very neatly. You can turn the molecule into boiled eggs with calcium and by taking calcium out with a chelating agent, it will turn into a jelly again.

Q: I am not sure quite how relevant this is, but Singer's done some recent work in which he has shown that actin and myosin elements accumulate directly under membrane patches, and he doesn't find involvement of microtubular structures as in some of the older models. This could provide an energy transduction mechanism from the membrane to the cytoplasmic region, and it might be quite sensitive to calcium too.

A: It is indeed. Hydén has shown that there is a fibrillar structure under the lipid bilayer of the cerebral neuron that is extremely calcium-sensitive in its fluorescence characteristics. Your question regarding connections to the microtubular apparatus, and thus the connections from the membrane to the nucleus that Edelman has been pushing so strongly, has been dealt some pretty severe blows in recent times. One such blow relates to a recent report in the Los Angeles Times, where the question was asked: "If you clone the nucleus of a cancer cell into a normal frog cell, what do you get — a knot of cancer cells or a tadpole?" The answer is that you get a tadpole. It really has dealt a very serious blow to the concept of the membrane as subservient to the nucleus. There are those of us who say, and with good reason, that many membranes in many circumstances are virtually autonomous and carry on a vast series of very complex transactions without the commitment to nuclear communication.

Q: It wasn't quite clear to me in your efflux studies whether you were dealing with extracellularly-bound calcium, or whether calcium fluxes were coming from the inside of the cell. I wonder if you could comment on this.

A: The question of where the calcium comes from in our efflux experiments is indeed a matter worth considering. There is typically about 2.4 millimolar calcium in the extracellular fluid, and it is down to 10^{-7} molar in the cytoplasm, excluding the organelles. Can calcium be mobilized from mitochondria and other places in the course of any excitatory process? The work of Mullins and others at the University of Maryland has shown that very little calcium appears to be mobilized from intracellular sites in the course of excitatory processes. Then most recently, Dr. Bawin has been studying the effects of lanthanum. Lanthanum is a membrane blocker that stops the movement of calcium in either direction across the membrane. There was a change in calcium efflux in Dr. Bawin's field experiment using lanthanum, which was attributed to extracellular calcium. The results of the field experiment were quantitatively different in ways that we need not discuss here, but which would appear to relate to a complexing between calcium and lanthanum. The point is that we have every reason to think that virtually all the calcium that is mobilized is on the surface of the cell.

Q: You were talking about an experiment in which you predicted which question was given to people on the basis of their EEG. You must be analyzing the spectral characteristics in some way. What was it in the EEG that permitted you to identify correctly the question asked to those people 90% of the time?

A: Actually, you have inverted the experimental situation. It was given that you knew what questions were asked, and you then had to classify the EEG records as coming from the period in which the subjects were considering that particular question. That is what could be done accurately in the hindsight of having asked a particular series of questions, since we found that you could take the mishmash of EEG record and classify it into categories. The things that were significant were primarily the levels of interrelationship, the mathematical coherence, between frontal-temporal and temporal-occipital areas. This was observed with frequencies that we know relate to the generation of EEG activity by deep structures in the temporal lobe, specifically the hippocampus. There are many other things that one can pull out of EEG records relating to correctness and incorrectness of decision-making. At the Langley-Porter Clinic, Dr. Enoch Calloway has done some really beautiful studies on scalp EEG's in men during decision-making, and classified at least five different types of decision-making processes based on the EEG record.

Q: You broadcast your low frequency signal as a modulation on an RF carrier signal, and this has been somehow detected by brain tissues. Since the waveform frequency really isn't in the Fourier analysis at all, the brain must be a nonlinear device similar to a detector in a radio. Is this correct, or does it obey Ohm's law?

A: This is quite an important question because one of the baffling aspects of our experimental scheme relates to the positioning of electrodes with one millimeter or so between them. In between the electrodes, there will be a population of, let's say, a couple of thousand cells and innumerable phase partitions. Every discontinuity represents a potential site for carrier rectification. If those domains in which rectification is occurring had any organized orientation with respect to the recording electrodes, there should be a net potential produced on the recording electrodes. We do not, in fact, observe such a potential. We have never seen it in many, many years of recording this type of activity. Nor do we see a gigantic rectification at the metal electrode interface at the moment that the field is turned on. What is suggested is that, at a much finer level of dimension, the field is being rectified by some change in local ionic concentration made at a rate corresponding to the field modulation. This could influence the separate behavior of neuronal cells, leading eventually to a condition in which, when the physiological inputs to these cells suggest that a particular rhythm signature should be generated, it is done in a facilitated fashion. Also, the rhythm signature is produced more often and for longer periods, so that you see higher densities in the spectrum. In my view, every phase partition in the tissue is a site of potential rectification.

Q: You showed that the ability to recognize or predict correctly a time interval is influenced by modulated RF. I would like to know what sort of neural circuitry is involved in making this recognition? Secondly, if you gave a modulated RF that is not several cycles per second but a different frequency, let us say slower than the time that you recognized, will that change the recognition in the other direction?

A: Yes, the selection of 7 Hz for the monkey experiments was deliberate because of our knowledge that the hippocampal system has a dominant 7-Hz frequency in the monkey and the suspicion that hippocampal function is involved in time estimations. That experiment was done with an ELF field and not with the modulated RF field. However, if one looks at the relative time that it takes for evidence of either the ELF field or the radiofrequency-modulated activities to appear in the EEG, one observes a very interesting sequence. For the ELF field, it will take two or three hours before you see increased 7-Hz activity in the hippocampus of the monkey. For the radiofrequency field, you start to see a generalized pickup of this activity after about three or four minutes. In some previous experiments with cats, it was shown that if you modulated the RF at 3 Hz, the animals would go into a slow-wave sleep and would stay in slow-wave sleep. This sleep was not like normal sleep, because it showed none of the remissions and none of the changes in state that normally occur. The cats would stay in one state for as long as five minutes. With a 16-Hz modulation on an RF signal, you could induce a type of REM sleep in cats that went on and on, so that the state dependence reflected the modulation characteristics in terms of the modulation frequency.

Chapter 7 # Theoretical Aspects of Magnetic Field Interactions With Biological Systems

Analysis of Stationary Magnetic Field Effects on Ionic Diffusion and Nerve Action Potentials

Richard L. Liboff

Magnetic fields could influence cellular development and function through an effect on the diffusion rate of ions across the plasma membrane.[1,2] However, a comparison of the Larmor radius of an ion with its mean free path in solution indicates that a stationary magnetic field of megagauss strength would be required to measurably affect diffusion. An experimental approach to this question has been taken, in which magnetically-induced changes in ion diffusion rates have been studied by measuring the conductivity of CsCl solutions in the presence and absence of a magnetic field. For this purpose, an ac bridge circuit was employed in which the frequency was $\approx 10^3$ times lower than the Larmor frequency ($\approx 10^7$ Hz at 1 kG). A null-point conductivity measurement was made to detect any imbalance of the bridge circuit resulting from application of a magnetic field, thus signifying an influence of the field on the ionic diffusion coefficient. No imbalance of the bridge circuit was observed following the application of fields of up to 1 kG strength, a result which is in conformity with theoretical predictions.

A possible inductive interaction between a steady magnetic field and a nerve action potential has also been investigated. The following analysis indicates that, because of the rotational symmetry of the current loops in

Richard L. Liboff, Schools of Applied Physics and Electrical Engineering, Cornell University, Ithaca, New York 14853.

the action potential, magnetic field effects tend to be nullified. The circulating current patterns that accompany the action potential have been described in the literature.[3] In the following analysis, an action potential propagating along a squid axon with a velocity $v \approx 10^3$ cm/sec[4,5] is assumed to enter a region with a steady magnetic field B directed transverse to the axis of the axon. It is further assumed that the leading edge of the current loops is approximately equal in length to the radius of the axon: $a \approx 250$ μm. The membrane conductance (G_m) and the axoplasm resistance (R_i) are approximately $2\pi a/2000$ mho · cm^{-1} and $200/\pi a^2$ ohm · cm^{-1}, respectively. Substituting these values into Faraday's law gives for the induced current in the axon:

$$I_{ind} \approx 4.5 \times 10^{-8} \alpha B,$$

where B is in gauss and α is a symmetry coefficient that varies from 0 to 1. If the action potential in the magnetic field maintains its cylindrical symmetry, $\alpha = 0$. For a severely asymmetric response, $\alpha \approx 1$.

The peak current density in the action potential of a squid axon is $J \approx 650$ μA·cm^{-2}. For a pulse 1 cm in length, $I \approx 51$ μA. Hence, for the current induced by a magnetic field to affect the action potential current, a field with a magnitude of approximately $1.1/\alpha$ kG and distributed over about 1 cm of axon would be required. It is clear that any field effect would tend to be cancelled by rotational symmetry of the action potential current flow, viz., $\alpha = 0$. Previous experimental studies indicated that no magnetic effects on the action potential of the frog sciatic nerve occurred at fields up to 100 kG distributed over 0.5 cm.[6]

The possibility of current distortion of the action potential due to a magnetic field was also investigated. The extent of such distortion was evaluated by comparing the Hall electric field, E_H, to the electric field, E_a, associated with the current flow of the action potential. For parameters appropriate to the squid axon, it was found that the ratio E_H/E_a remains less than unity for magnetic fields less than ≈ 1 kG.

REFERENCES

1. Liboff, R. L., 1965. A biomagnetic hypothesis. *Biophys. J.* **5**:845-853.
2. Liboff, R. L., 1969. Biomagnetic hypotheses. In *Biological Effects of Magnetic Fields*. Vol. 2, ed., M. F. Barnothy. New York: Plenum Press.
3. Keynes, R. D., 1958. The nerve impulse and the squid. *Sci. Amer.* **199**(6):83-90.
4. Hodgkin, A. L. and Huxley, A. F., 1952. A quantitative description of membrane current and its application to conduction and excitation in nerve. *J. Physiol.* **117**:500-544.
5. Katz, B., 1966. *Nerve, muscle and synapse*. New York: McGraw-Hill Book Company.
6. Liberman, E. A., Vaintsvaig, M. N., and Tsofina, L. M., 1959. The effect of a constant magnetic field on the excitation threshold of isolated frog nerve. *Biophysics* **4**:152-155.

DISCUSSION

Q: I would like to point out an interesting phenomenon that could occur in the nervous system when there are nerves entwined around each other. If the nerves are sufficiently close together, one would expect that the passage of current down one nerve would establish an electric field and a current flow in the nerves with which it is intertwined. In other words, there would be a type of non-synaptic synapse. There have not been any experimental tests of this phenomenon, but the snail would be an ideal organism for such studies since it has a large number of entwined nerves.

A: That is a very interesting proposal, and it would be worthwhile to see if such situations might exist in the brain.

Mechanisms of Biomagnetic Effects

D. Dennis Mahlum

The development stages of many biological systems are extremely sensitive to a number of physical and chemical insults, resulting from the multiplicity of interactive processes required to attain normal development. Our studies on these development stages involve the use of two parallel animal test systems, the mouse and the trout, to obtain comparative data for warm- and cold-blooded organisms.

The mouse studies are being performed using a replicate experimental design that permits prolonged exposures of mice in different stages of gestation to static homogeneous (10,000 G), static gradient (250 G/cm), or cyclic gradient fields (250 G/cm, approximately 0.6 cycle/min). Some animals are examined prenatally for morphologic effects, while others are evaluated postnatally for functional changes.

In the trout studies, eggs are fertilized and permitted to develop in the magnetic field for 21 days (approximately 7 days before hatching). The eggs are then removed from the field and allowed to hatch in a flow-through incubator. Several endpoints, including morphologic alterations, hatching efficiency, and survival are used to assess the effects of the exposure.

These studies should provide an integrated assessment of biomagnetic effects in a manner that is independent both of the level at which the effect occurs, and of the mechanism. In addition, we are investigating a number of simpler *in vitro* systems including: (a) changes in permeability characteristics of artificial and natural membranes, (b) changes in gelation temperatures of

D. Dennis Mahlum, Department of Biology. Battelle-Pacific Northwest Laboratories, Richland, Washington 99352.

macromolecules, (c) effects on growth and morphologic and functional characteristics of cells in tissue culture, and (d) effects on neuromuscular function. Studies to search for genetic effects using microbial and animal systems are also in progress.

At a minimum, the information being produced from these studies will provide a data base for establishing exposure standards. Moreover, if reproducible effects are found, they will provide a system for mechanistic and dosimetric studies. Current speculation in our laboratory suggests that alterations in macromolecular or membrane conformation could result in biochemical and physiological changes which would serve to amplify small effects produced by interaction of magnetic fields with biological material. Singer's fluid mosaic model of membrane structure may provide a tool for exploring the validity of these speculations, should reproducible effects be identified.

DISCUSSION

Q: You described some experiments which indicate that magnetic field exposure leads to an increase in the temperature at which gelation occurs. Are your gels diamagnetic, and what do you propose to be the mechanism underlying the change in gelation temperature in the presence of a magnetic field?

A: The gels are diamagnetic. We feel that the magnetic field is probably affecting alignment of molecules in the gel, and thereby influencing the gelation temperature.

Magnetic Field Coupling With Liquid Crystalline Structures

Mortimer M. Labes

Liquid crystalline phase of two types have been identified in biological systems: "lyotropic" phases, which are assemblies of oriented rod-like solutes in a solvent, and "thermotropic" phases, which consist of oriented rod-like molecules.

In a magnetic field, arrays of diamagnetic rods adopt a position which minimizes their free energy. If the diamagnetic anisotropy $X_a = X_{11} - X\perp$ is > 0, then the long molecular axis aligns in the field. The free energy perturbation is $F_{magnetic} = - 1/2 X_a (\hat{n}H)^2$, where $\hat{n}$ is the director of the liquid crystal and H is the magnetic field strength. The critical magnetic fields to produce reorientation of the director in a thin sample (~ 0.1 mm in depth) is typically $10^3 - 10^4$ Oe (depending on the magnitude of X_a). Subcritical fields perturb cholesteric liquid crystals by changing the pitch of the cholesteric helix.

Transport processes are known to be anisotropic in liquid crystals. For example, diffusion coefficients are typically twice as large parallel to the long axis as compared to the perpendicular direction. The proposed model for biomagnetic effects consists, therefore, of the perturbation of a transport process through a liquid crystalline membrane. The perturbation can be expected to influence the rate of transport by a factor of 2 for magnetic fields of $10^3 - 10^4$ Oe.

This model was first published in 1966.[1] Measurements of magnetic field effects on diffusion in liquid crystals,[2] and on the anisotropy of diffusion have also appeared.[3,4] These reports confirm several aspects of the above model of biomagnetic effects.

REFERENCES

1. Labes, M. M., 1966. A possible explanation for the effect of magnetic fields on biological systems. *Nature* **211**:968.
2. Teucher, I., Baessler, H., and Labes, M. M., 1971. Diffusion through nematic liquid crystals. *Nature (London) Phys. Sci.* **229**:25-26.
3. Hakemi, H., and Labes, M. M., 1974. New optical method for studying anisotropic diffusion in liquid crystals. *J. Chem. Phys.* **61**:4020-4025.
4. Hakemi, H., and Labes, M. M., 1975. Self-diffusion coefficients of a nematic liquid crystal via an optical method. *J. Chem. Phys.* **63**:3708-3712.

Mortimer M. Labes, Department of Chemistry, Temple University, Philadelphia, Pennsylvania 19122.

DISCUSSION

Q: I would be interested in your opinion regarding ways in which the magnetic field phenomena that you describe might influence phase changes in membranes.

A: A magnetic field would be expected to affect structural features in liquid crystalline portions of the membrane, particularly in certain ranges of temperature. This type of effect can be demonstrated quite clearly with cholesteric liquid crystals that change their pitch as a function of temperature. The magnetic field effect becomes very pronounced in certain discrete temperature ranges. For example, some cholesteric systems change from a right-handed to a left-handed pitch at a well-defined temperature. Near this transition point, a strong magnetic field can profoundly influence structure when the pitch is small and molecular packing is rather loose. Under this condition, the magnetic field can set up cooperative interactions that have a large effect on long-range ordering within the liquid crystal. This type of effect can be detected by a perturbation in the dielectric anisotropy of the liquid crystal in the presence of a magnetic field. Thus we have a situation where structure variations as a function of temperature can influence the magnetic field effect. Similar situations may exist in biomembranes.

Q: Does the magnetic field shift the transition temperature?

A: In the case of the cholesteric liquid crystals that I just discussed, there does not appear to be a shift in the transition temperature in the presence of a magnetic field. In fact, one would not expect such a change on theoretical grounds, and I do not know of any experimental situation with liquid crystals where magnetic fields have been observed to shift the phase transition temperature by more than a few tenths of a degree.

Q: Although the transition temperature may not be influenced, is it not the case that diffusion and transport processes may be more severely perturbed by the application of a magnetic field at temperatures close to the transition point?

A: Yes, there is less anisotropy of diffusion as the liquid crystal undergoes a phase transition. The apparent diffusivity drops to a value which is the average of the values along directions that are parallel and perpendicular to the axis of the liquid crystal, and this can represent a change in the magnitude of the diffusivity of about 1.5. Thus, at the phase transition, transport properties are changing rapidly and these changes can be further amplified by applying a magnetic field.

Superconductive Josephson Junctions — A Possible Mechanism for Detection of Weak Magnetic Fields and of Microwaves by Living Organisms

Freeman W. Cope

The extreme sensitivity of several organisms to very weak magnetic fields (~0.1 to 5 G) has led to the proposal that superconductive junction phenomena may occur in these biological systems. The form of the temperature dependence of various nerve and growth processes is also consistent with the suggestion that electron tunneling between superconductive microregions might be rate limiting for these physiological processes.[1] Although no direct evidence for this phenomenon has been obtained *in vivo*, various model chemical systems have been examined for the presence of superconductive transitions and Josephson junctions at temperatures in the physiological range. For example, experimental evidence is available for probable superconductive transitions in the cholates at close to room temperature.[2,3] Also, weak magnetic field effects on probable Josephson junctions have been observed in carbon films at $25°C$.[4] Electron tunneling through these junctions responds to very small magnetic fields, as evidenced by the fact that superconducting Josephson junctions have been fabricated which exhibit sensitivity to fields as small as 10^{-11} G. Josephson junctions between superconductive microregions in living systems may thus provide a possible physical mechanism with more than enough sensitivity to explain the observed responses of organisms to weak magnetic fields.[5] In addition, microwave effects observed on probable room temperature Josephson junctions in carbon films[6] suggest that non-thermal effects of microwaves in biological systems may occur with high sensitivity by the same mechanism.[7]

REFERENCES

1. Cope, F. W., 1971. Evidence from activation energies for superconductive tunneling in biological systems at physiological temperatures. *Physiol. Chem. Phys.* 3:403-410.
2. Wolf, A. A., and Halpern, E. H., 1976. On a class of organic superconductors: A summary of findings. *Proc. IEEE* **64**:357-359.
3. Wolf, A. A., 1976. Experimental evidence for high-temperature organic fractional superconduction in cholates. *Physiol. Chem. Phys.* **8**(6):495-518.
4. Antonowicz, K., 1974. Possible superconductivity at room temperature. *Nature (London)* **247**:358-360.

Freeman W. Cope, Biochemistry Laboratory, Naval Air Development Center, Warminster, Pennsylvania 18974.

5. Cope, F. W., 1973. Biological sensitivity to weak magnetic fields due to biological superconductive Josephson junctions? *Physiol. Chem. Phys.* **5**:173-176.

6. Antonowicz, K., 1975. The effect of microwaves on dc current in an Al-carbon-Al sandwich. *Phys. Status Solidi* **a28**:497-502.

7. Cope, F. W., 1976. Superconductivity − a possible mechanism for non-thermal biological effects of microwaves. *J. Microwave Power* **11**:267-270.

DISCUSSION

Q: There are several well-known charge transfer mechanisms that operate in biological systems. For example, short-range electron transfer occurs between molecules in membranes. There is also longer range charge transfer by ions. Could you clarify why it is necessary to propose the existence of superconductivity in view of these alternative mechanisms that have been demonstrated experimentally?

A: The charge transfer mechanisms that you named do not appear to have the sensitivity to low magnetic fields that we know exists in a number of biological systems. In general, electron and ion charge transfer processes are liquid state phenomena and are influenced only by fields in excess of one kilogauss. Perhaps there are special conditions that would lead to a greater sensitivity, but I do not think that these have been observed as yet. In contrast, it is known that solid state superconductive junctions are sensitive to fields on the order of 1 G or less, and that is why I have proposed that they exist in biological systems that are influenced by very low fields.

Q: What is a cholate and how does it function as a superconductor?

A: It is a ring structure with several side groups, and there is nothing obvious about this class of molecules that should make them uniquely qualified as superconductors. It just happened that Wolf discovered this property when he was screening a number of organic compounds for the property of high temperature superconductivity. It is interesting to note that some water must be present in order to observe superconduction in the cholates. This fact suggests that they may be the "sandwich" type of superconducting structure proposed by Ginsburg, in which a thin film conductor is adjacent to a strong dielectric so that electrons can propagate in the conductor while phonons propagate in the adjacent dielectric. Actually, biological systems are full of such sandwich structures in which layers of water are interfaced with protein and lipid charge conductors.

Theoretical Remarks on Low Magnetic Field Interactions With Biological Systems

Charles E. Swenberg

The influence of external magnetic fields on chemical and biological systems may be classified as either "static" or "dynamic" effects. Static effects arise from classical diamagnetic, ferromagnetic or paramagnetic interactions with the applied field, and generally result from the long-range structural organization of the responsive component of the system. In contrast, dynamic magnetic processes involve only a few reactant molecules — generally two — and the reaction is cooperative in the sense that the reactants enter an intermediate state with a characteristic coherence time, τ. For values of τ greater than a critical value, on the order of one nanosecond, reactions that involve radical pairs as intermediates are sensitive to the application of small magnetic fields (<1 kG).

A number of examples of dynamic magnetic processes exist in organic solids and liquids.[1] All of these processes have a characteristic dependence on the applied magnetic field, and exhibit saturation above a certain field strength, H_S. In systems where $H_S \approx 100$ G, the field modulation of the reaction owes its origin to hyperfine interactions. Examples of such processes include dye-induced charge injection into an organic solid[2,3] and electron transfer reactions in polar solvents.[4,5] For reactions that involve paramagnetic bimolecular reaction kinetics, the saturation field $H_S \approx 1$ kG. In this case, fine structure and Zeeman interactions are dominant. Examples of such systems include triplet-triplet exciton annihilation, electroluminescence, radiation-induced scintillation from organic solids, and electrogenerated chemiluminescence.

An example of biomagnetic hyperfine interactions is provided by the photochemical properties of the bacterium *Rhodopseudomonas sphaeroides*.[6,7] In this system, if the photochemical electron transfer to the Fe-ubiquinone complex is blocked by chemical reduction, then the lifetime of the radical pair intermediate ($P870^+I^-$)formed by a flash excitation will increase from 0.2 to 10 nanoseconds. Under this condition, a magnetic field effect on the spin state of the intermediate can be observed as a change in the absorption spectrum of the carotenoid receptor. The saturation field is approximately 100 G, and the maximum magnetic field on the absorption is about 30%.

Other examples of biomagnetic hyperfine interactions may arise in systems where a photoisomerization is coupled to an exothermic transfer reaction. It has been proposed that the sensitivity of certain migratory birds

Charles E. Swenberg, Radiation and Solid State Laboratory, New York University, New York, New York 10003.

to the geomagnetic field may originate from an anisotropic hyperfine tunneling mechanism that serves as an internal compass.[8] In cases where light is necessary for detection of both the presence and direction of the earth's field, small fluctuations are predicted to occur in the plasma membrane current of retinal photoreceptors cells when they have different spatial orientations relative to the field.

REFERENCES

1. Swenberg, C. E. and Geacintov, N. E., 1973. *Organic molecular photophysics*, Ch. 10, ed., J. B. Birks. New York: John Wiley and Sons, Inc.
2. Groff, R. P., Merrifield, R. E., Suna, A., and Avakian, P., 1972. Magnetic hyperfine modulation of dye-sensitized delayed fluorescence in an organic crystal. *Phys. Rev. Lett.* **29**:429-431.
3. Groff, R. P., Suna, A., Avakian, P., and Merrifield, R. E., 1974. Magnetic hyperfine modulation of dye-sensitized delayed fluorescence in organic crystals. *Phys. Rev.* **B9**:2655-2660.
4. Schulten, K. and Weller, A., 1978. Exploring fast electron transfer processes by magnetic fields. *Biophys. J.* **24**(1):295-305.
5. Schulten, K., Staerk, H., Weller, A., Werner, H. J., and Nickel, B., 1977. Magnetic field dependence of the germinate recombination of radical ion pairs in polar solvents. *Z. Phys. Chem.* **NF101**:371-390.
6. Blankenship, R. E., Schaafsma, T. J., and Parson, W. W., 1977. Magnetic field effects on radical pair intermediates in bacterial photosynthesis. *Biochim. Biophys. Acta.* **461**:297-305.
7. Hoff, A. J., Rademaker, H., Grondell, R. V., and Duysens, L. N. M., 1977. On the magnetic field dependence of the yield of the triplet state in reaction centers of photosynthetic bacteria. *Biochim. Biophys. Acta* **460**:547-554.
8. Schulten, K., Swenberg, C. E., and Weller, A., in press. A biomagnetic sensory mechanism based on magnetic field modulated coherent electron spin motion. *Z. Physik. Chem.*

DISCUSSION

Q: In the spin-conserving reactions that you considered, are you saying that a low magnetic field decouples the electron and nuclear spins of the reactants and therefore suppresses the triplet channel?

A: That is correct.

Q: If you go to very high fields, is there any effect on the matrix element that might connect the singlet state with the $m_S = 0$ of the triplet state? In other words, could an effect occur at extremely high fields that causes the reaction to go in the opposite direction to that observed in lower fields?

A: When there is no applied field, the singlet state is coupled to all states of the triplet manifolds. At high magnetic fields, the splitting of triplet manifolds decouples the plus and minus states from the singlet, and therefore you have fewer channels. This suppression of the number of channels decreases the overall rate constant for the particular reaction being considered.

Q: In discussing charge transfer reactions, you stated that the tunneling distance for a proton is on the order of a few angstroms. It seems to me that there is no reason to think that there is a limit on the tunneling distance, since it depends on the characteristic dimensions of the barrier.

A: As far as I know, the maximum tunneling distances in processes that have been fully analyzed are between 6 and 10 angstroms. This could place a severe restriction on biological transport processes that rely on tunneling phenomena. It would also confer directionality in terms of structure as well as chemistry, and this might be important in biological systems. If the tunneling distance were long, say 50 angstroms, then the charge transfer process would involve large numbers of molecules and could proceed in too many directions.

Chapter 8 Summary

Tom S. Tenforde

A major objective of the Biomagnetic Effects Workshop was to review the current state of knowledge in the field of biomagnetics, and this goal has been accomplished reasonably well. Several of the major points made in this conference, as I see them, were the following:

1. *Lower Organisms.* The magnetic field sensitivity exhibited by magnetotactic bacteria, sharks and rays, migratory birds, and honeybees were reviewed by Dr. Blakemore, Dr. Kalmijn, and Dr. Keeton. The sensitivity of birds and insects to low frequency electromagnetic fields was also described by Dr. Keeton and Dr. Greenberg.

Sensitivity of the various strains of magnetotactic bacteria to fields of geomagnetic strength results from the presence of a large quantity of ferromagnetic iron. In the shark, motion through the earth's field induces an EMF in special electroreceptor organs, the ampullae of Lorenzini. Fields as small as 0.01 μV/cm can be detected and used as orientational cues. Some new and intriguing ideas regarding the amplification of such weak signals through cooperative phenomena occurring at the surfaces of brain cells were described by Dr. Adey. The mechanism(s) by which birds and insects sense fields on the order of 1 G is totally unknown. A unique hypothesis was put forth by Dr. Swenberg involving the effects of small fields on photoreception, leading to the concept of a "visual compass."

Given our present knowledge, it is difficult to make generalizations regarding the sensitivity of lower organisms to very weak magnetic and electromagnetic fields. Observations made to date suggest that a diversity of sensory mechanisms may have evolved independently for a wide variety of phylogenetically unrelated organisms.

Tom S. Tenforde, Biology and Medicine Division, Lawrence Berkeley Laboratory, Berkeley, California 94720.

2. *Mammals.* A large number of alterations in mammalian physiology and behavior has been reported to occur following exposure to stationary and low-frequency magnetic fields. These were reviewed by Dr. Sheppard, who pointed out the rather confused picture that has emerged because of the wide variety of experimental conditions and magnetic field parameters that have been used in different laboratories. Another aspect of this problem is the frequent lack of reproducibility of results obtained by different investigators. For example, several published reports of magnetic field effects on hematological parameters, tissue histopathology and behavioral patterns in mammals must be regarded as questionable in view of the negative findings described at this Workshop by Dr. Biggs, Dr. Nahas and Dr. de Lorge.

At present, it would seem that a concerted effort is required to amass data on a broad range of mammalian systems under magnetic field exposure conditions that are free of experimental artifacts. An immediate impetus for such studies is provided by the requirement to establish exposure guidelines for workers in several newly developing technologies, as described by Dr. Alpen. In response to this need for biomagnetic studies on mammalian systems, a programmatic effort described by Dr. Mahlum is being carried out at Battelle-Pacific Northwest Laboratories in Washington. A parallel program involving several whole-animal, tissue, and organ systems is also under way at the Lawrence Berkeley Laboratory.

3. *Cellular and molecular systems.* The growth rate, reproductive integrity and radiation response of cultured mammalian cells was reported by Dr. Rockwell to be unaffected by fields as high as 20 kG. Similarly, Dr. Weissbluth presented evidence that exposures to fields up to 220 kG had no effect on the reaction rates of several enzymes that are important for cellular function. Both the theoretical and experimental aspects of magnetic field interactions with ionic currents in nerve axons were reviewed by Dr. Liboff, who concluded that the rotational symmetry of the current loops tends to cancel any magnetic effect on the propagation of action potentials. No obvious cytogenetic changes were apparent from the experiments reported by Dr. Baum, in which he examined the rate of mutation in eukaryotic plant cells following a prolonged exposure to fields as high as 37 kG.

In contrast, Dr. Malinin reported that large stationary magnetic fields led to inhibition of DNA and hemoglobin synthesis and caused morphological transformation when applied to cells in the frozen state. The implication of these results for cellular function when the magnetic field exposure occurs under physiological conditions is presently unclear. An inhibitory effect of a 5 to 9 kG field on the contractile mechanism in a protozoan organism was reported by Dr. Ettienne, which he attributed to a decreased rate of enzymatic transport of calcium ions out of the cell cytoplasm following contraction.

The orientational effects of stationary magnetic fields on retinal rods, chloroplasts and intact *Chlorella* cells were reviewed by Dr. Hong and Dr. Geacintov. There is evidence that such orientational effects, which are observed in fields of approximately 10 kG strength, originate because of an anisotropy in the diamagnetic susceptibility of oriented membrane components. Dr. Labes described the effects of magnetically-induced orientation on transport processes in liquid crystalline membranes.

One target that has not been directly explored for magnetic field effects, and which could have a profound influence on cellular functions, is the ferricytochrome-rich electron transport chain. The cytochromes may be ordered over a large spatial domain, and could exhibit paramagnetic orientation in strong fields. The influence of magnetic fields on cell metabolism and specialized functions that depend upon the respiratory chain deserve consideration in future biomagnetic studies.

4. *Theory.* Except for orientational effects on magnetically anisotropic macromolecular structures, very few biomagnetic phenomena have been interpreted in well-defined theoretical terms. Some relatively new concepts that were discussed at the Workshop concerned the sensitivity of certain organisms to very small magnetic fields. Dr. Cope proposed that such weak fields could influence superconductive tunneling phenomena in living cells. At present, however, there does not appear to be direct evidence for the existence of superconductivity in biological systems. A field-sensing mechanism involving an anisotropic hyperfine interaction in the radical-pair intermediates of electron transfer reactions was proposed by Dr. Swenberg. Further experimental tests will be required to determine the applicability of such mechanisms to observed biomagnetic effects.

5. *Medical applications.* A new technique for modifying the depth-dose profile of fast electron beams used in radiotherapy was described by Dr. Nath. By applying a transverse magnetic field, it has been possible to enhance the dose delivered in a treatment zone near the end of the electron range relative to the entrance dose. The resulting dose versus depth characteristics resemble those for negative pions and heavy ions, and may prove advantageous in the therapy of deep-seated tumors.

In concluding, I would like to thank all of the participants at the Workshop for their carefully prepared presentations. Their efforts have served to outline both the advances that have been made in the past, and the areas that will require intensive research in future biomagnetic studies.

Appendix I

Biomagnetic Effects Workshop Program

Sponsoring Organization: Division of Biology and Medicine, Lawrence Berkeley Laboratory, University of California at Berkeley

Dates: April 6-7, 1978

Location: Building 50, Auditorium, Lawrence Berkeley Laboratory

Registration: (For all conference speakers and attendees) 8:00-9:00 a.m. on April 6 in the lobby of Auditorium, Building 50.

PROGRAM
Thursday, April 6

9:00 a.m.   T. Tenforde, Lawrence Berkeley Laboratory: Introductory Remarks

9:20 a.m.   Panel Discussion: Effects of Magnetic Fields on Lower Organisms
Participants:
- R. Blakemore, Department of Microbiology, University of New Hampshire
- Ad. J. Kalmijn, Woods Hole Oceanographic Institution
- W. T. Keeton, Section of Neurobiology and Behavior, Cornell University
- B. Greenberg, Department of Biological Sciences, University of Illinois, Chicago Circle
- J. Baum, Safety and Environmental Protection Division, Brookhaven National Laboratory

(Coffee break: 10:15-10:30)

12:15 p.m.   Lunch (no-host) in LBL Cafeteria

1:15 p.m.   E. Alpen, Lawrence Berkeley Laboratory: Magnetic Field Exposure Guidelines

1:35 p.m.   Panel Discussion: Effects of Magnetic Fields on Mammals
             Participants:
             A. R. Sheppard, Jerry L. Pettis Memorial V.A.
                Hospital, Loma Linda, California
             J. de Lorge, Naval Aerospace Medical Research
                Laboratory, Pensacola
             G. G. Nahas, College of Physicians and Surgeons,
                Columbia University
             M. W. Biggs, Department of Industrial Medicine,
                Lawrence Livermore Laboratory

             (Coffee break: 3:15-3:30)

3:30 p.m.   Panel Discussion: Magnetic Field Effects on Cellular and
             Molecular Systems
             Participants:
             F. T. Hong, Department of Physiology, Wayne State
                University, School of Medicine
             N. E. Geacintov, Department of Chemistry, New York
                University
             M. Weissbluth, Department of Applied Physics,
                Stanford University
             G. I. Malinin, Department of Physics, Georgetown
                University
             S. Rockwell and R. Nath, Department of Therapeutic
                Radiology, Yale University
             E. Ettienne, Department of Physiology, University of
                Massachusetts Medical School

6:15-7:30 p.m. Cocktails (no-host) at the University of California Faculty
                Club

7:30 p.m.   Dinner in the Great Hall of the University of California
             Faculty Club

PROGRAM
Friday, April 7

8:30 a.m.   W. R. Adey, Jerry L. Pettis V. A. Hospital, Loma Linda,
             California: Long-Range Electromagnetic Field
             Interactions at Brain Cell Surfaces

9:20 a.m.   Panel Discussion: Mechanisms of Magnetic Field
             Interactions With Biological Systems
             Participants:
             R. L. Liboff, Schools of Applied Physics and Electrical
                Engineering, Cornell University
             D. D. Mahlum, Department of Biology, Battelle—

                     Pacific Northwest Labs

                     M. M. Labes, Department of Chemistry,
                     Temple University

                     F. W. Cope, Naval Air Development Center,
                     Warminster, Pennsylvania

                     C. E. Swenberg, Radiation and Solid State
                     Laboratory, New York University

             (Coffee break: 10:15-10:30)

12:15 p.m.   Lunch (no-host) in LBL Cafeteria

1:30 p.m.   Tour of LBL facilities: Bevatron, 184-Inch Cyclotron,
             Magnetic Effects Laboratory

             DOE Committee on Magnetic Field Exposure Guidelines
             (Chairman: E. Alpen) will meet in executive session at the
             Donner Laboratory, Room 458.

Appendix II

Workshop Participants

Speakers

W. Ross Adey
Research Service 151, Jerry L. Pettis Memorial V. A. Hospital, Loma Linda, CA 92357

Edward L. Alpen
Donner Laboratory, Room 466, Lawrence Berkeley Laboratory, University of California, Berkeley, CA 94720

John Baum
Safety and Environmental Protection Division, Building 535, Brookhaven National Laboratory, Upton, NY 11973

Max W. Biggs
Department of Industrial Medicine, Lawrence Livermore Laboratory, P.O. Box 808, L-L423, Livermore, CA 94550

Richard Blakemore
Department of Microbiology, University of New Hampshire, Spaulding Life Sciences Building, Durham, NH 03824

Freeman W. Cope
Biochemistry Laboratory, Code 6022, Naval Air Development Center, Warminster, PA 18974

John de Lorge
Naval Aerospace Medical Research Laboratory, L33 JD, Pensacola, FL 32508

Earl Ettienne
Department of Physiology, University of Massachusetts Medical School, 55 Lake Avenue North, Worcester, MA 01605

Nicholas E. Geacintov
Department of Chemistry, Room 507, New York University, 4 Washington Place, New York, NY 10003

Bernard Greenberg
Department of Biological Sciences, University of Illinois at Chicago Circle, Chicago, IL 60680

Felix T. Hong
Department of Physiology, Wayne State University, School of Medicine, 540 E. Canfield Avenue, Detroit, MI 48201

Ad. J. Kalmijn
Department of Biology, Room 411, Woods Hole Oceanographic Institution, Woods Hole, MA 02543

William T. Keeton
Section of Neurobiology and Behavior, 141 Langmuir, Cornell University, Ithaca, NY 14853

Mortimer M. Labes
Department of Chemistry, Temple University, Philadelphia, PA 19122

Richard L. Liboff
Schools of Applied Physics and Electrical Engineering, Phillips Hall, Cornell University, Ithaca, NY 14853

D. Dennis Mahlum
Department of Biology, Battelle—Pacific Northwest Laboratories, Richland, WA 99352

George I. Malinin
Department of Physics, Georgetown University, 37th and "O" Streets, N.W., Washington, DC 20057

Gabriel G. Nahas
Departments of Anesthesiology and Pathology, College of Physicians and Surgeons, Columbia University, 630 West 168th Street, New York, NY 10032

Ravinder Nath
Department of Therapeutic Radiology, Yale University, School of Medicine, 333 Cedar Street, New Haven CT 06510

Sara Rockwell
   Department of Therapeutic Radiology, Yale University, School of
   Medicine, 333 Cedar Street, New Haven, CT 06510

Asher R. Sheppard
   Research Service 151, Jerry L. Pettis Memorial V. A. Hospital,
   Loma Linda, CA 92357

Charles E. Swenberg
   Radiation and Solid State Laboratory, Room 812, New York
   University, 4 Washington Place, New York, NY 10003

Tom S. Tenforde
   Biology and Medicine Division, Building 74, Room 344, Lawrence
   Berkeley Laboratory, University of California, Berkeley, CA 94720

Mitchell Weissbluth
   Department of Applied Physics, Stanford University, Stanford,
   CA 94305

Attendees

K. R. Baker
   Department of Energy, Washington, DC 20545

Bernard Baratz
   Environmental Technology Division, Office of Environmental
   Activities, D.O.E., Room 4220, 20 Massachusetts Avenue, N.W.,
   Washington, DC 20545

Edward L. Bennett
   Building 3, Room 202, Lawrence Berkeley Laboratory, University of
   California, Berkeley, CA 94720

Jacques Breton
   Chemistry Department, 215 Hildebrand Hall, University of California,
   Berkeley, CA 94720

Thomas F. Budinger
   Building 1, Room 230, Lawrence Berkeley Laboratory, University of
   California, Berkeley, CA 94720

Wing Chan
   Division of Allied Health, School of Medicine, 1100 West Michigan
   Street, Indiana/Purdue Medical Center, Indianapolis, IN 46202

George T. Chen
 Building 55, Room 123, Lawrence Berkeley Laboratory, University of California, Berkeley, CA 94720

John S. Colonias
 Building B50A, Room 101C, Lawrence Berkeley Laboratory, University of California, Berkeley, CA 94720

Stanley B. Curtis
 Building 74, Room 159B, Lawrence Berkeley Laboratory, University of California, Berkeley, CA 94720

Charlie Damm
 Lawrence Livermore Laboratory, P.O. Box 808, L-441, Livermore, CA 94550

Tom Distler
 Lawrence Livermore Laboratory, P.O. Box 808, L-386, Livermore, CA 94550

Charles G. Dols
 Building 70, Room 261, Lawrence Berkeley Laboratory, University of California, Berkeley, CA 94720

Shirley Ebbe
 Building 1, Room 227, Lawrence Berkeley Laboratory, University of California, Berkeley, CA 94720

Charles Ehret
 Argonne National Laboratory, 9700 S. Cass Avenue, Argonne, IL 60439

Merrill Eisenbud
 Institute of Environmental Medicine, New York University, Box 817, Tuxedo, NY 10987

Jacob Fabrikant
 Building 1, Room 468, Lawrence Berkeley Laboratory, University of California, Berkeley, CA 94720

Richard B. Frankel
 Francis Bitter National Magnet Laboratory, Massachusetts Institute of Technology, Cambridge, MA 02139

Cornelius Gaffey
 Building 74, Room 385A, Lawrence Berkeley Laboratory, University of California, Berkeley, CA 94720

Peter Gibbs
Physics Department, University of California, Berkeley, CA 94720

Murlin F. Gillis
Biology Department, Battelle—Pacific Northwest Laboratories,
Richland, WA 99352

William D. Gregory
Department of Physics, Georgetown University, 37th and "O" Streets,
N.W., Washington, DC 20057

Thomas L. Hayes
Building 1, Room 108A, Lawrence Berkeley Laboratory, University of
California, Berkeley, CA 94720

William R. Holley
Building 88, Room 115, Lawrence Berkeley Laboratory, University of
California, Berkeley, CA 94720

Jerry Howard
Building 51, Room 208, Lawrence Berkeley Laboratory, University of
California, Berkeley, CA 94720

Steve Hurst
Building 74, Room 344, Lawrence Berkeley Laboratory, University of
California, Berkeley, CA 94720

Marge Hutchinson
Building 50, Room 306, Lawrence Berkeley Laboratory, University of
California, Berkeley, CA 94720

Leal L. Kanstein
Building 80, Room 109, Lawrence Berkeley Laboratory, University of
California, Berkeley, CA 94720

Christine Laszcz-Davis
Lawrence Livermore Laboratory, P.O. Box 808, L-384, Livermore,
CA 94550

Robert Latimer
Building 4, Room 203B, Lawrence Berkeley Laboratory, University of
California, Berkeley, CA 94720.

John A. Linfoot
Building 57, Room 420C, Lawrence Berkeley Laboratory, University of
California, Berkeley, CA 94720

J. T. Lyman
   Building 55, Room 125, Lawrence Berkeley Laboratory, University of California, Berkeley, CA 94720

John L. Magee
   Building 29, Room 206, Lawrence Berkeley Laboratory, University of California, Berkeley, CA 94720

Michael Malachowski
   Building 10, Lawrence Berkeley Laboratory, University of California, Berkeley, CA 94720

Tod H. Mikuriya
   Psychosomatic Medicine, Gladman Center, Claremont Hotel, 41 Tunnel Road, Berkeley, CA 94705

Donald H. Nelson
   Building 46, Room 132, Lawrence Berkeley Laboratory, University of California, Berkeley, CA 94720

Frank Q. Ngo
   Building 29, Room 210, Lawrence Berkeley Laboratory, University of California, Berkeley, CA 94720

Michael S. Raybourn
   Building 1, Room 359, Lawrence Berkeley Laboratory, University of California, Berkeley, CA 94720

Sam Rogers
   Montana State University, Bozeman, MT 59715

Ruth J. Roots
   Building 10, Room 202, Lawrence Berkeley Laboratory, University of California, Berkeley, CA 94720

Keith A. Rose
   Building 1, Room 304, Lawrence Berkeley Laboratory, University of California, Berkeley, CA 94720

William Ross
   Lawrence Livermore Laboratory, P.O. Box 808, L-443, Livermore, CA 94550

G. J. Rotariu
   Department of Energy, Washington, D.C. 20545

Gwen J. Ryan
Building 74, Room 344, Lawrence Berkeley Laboratory, University of California, Berkeley, CA 94720

Walter Schimmerling
Building 29, Room 215C, Lawrence Berkeley Laboratory, University of California, Berkeley, CA 94720

John C. Schooley
Building 74, Room 319B, Lawrence Berkeley Laboratory, University of California, Berkeley, CA 94720

Melvin R. Sikov
Battelle–Pacific Northwest Laboratories, Richland, WA 99352

Helene S. Smith
Naval Biological Laboratory, T-19, University of California, Berkeley, CA 94720

Ralph H. Thomas
Building 72, Room 124, Lawrence Berkeley Laboratory, University of California, Berkeley, CA 94720

R. A. Tobey
Los Alamos Scientific Laboratory, University of California, P.O. Box 1663, Los Alamos, NM 87544

Cornelius A. Tobias
Building 10, Room 202, Lawrence Berkeley Laboratory, University of California, Berkeley, CA 94720

Frank T. Upham
Building 1, Room 325, Lawrence Berkeley Laboratory, University of California, Berkeley, CA 94720

William J. Vaughan
Building 74, Room 166, Lawrence Berkeley Laboratory, University of California, Berkeley, CA 94720

J. R. Wayland
(Sandia Labs), Los Alamos Scientific Laboratory, University of California, P.O. Box 1663, Los Alamos, NM 87544

Graeme P. Welch
Building 29, Room 224, Lawrence Berkeley Laboratory, University of California, Berkeley, CA 94720

Louis Wendt
Chemistry Station, Montana State University, Bozeman, MT 59715

Priscilla D. Wong
Building 1, Room 224, Lawrence Berkeley Laboratory, University of California, Berkeley, CA 94720

Manley Wu
Department of Energy, Environment and Safety Division, 1333 Broadway, Oakland, CA 94612

Tracy C. Yang
Building 74B, Room 106, Lawrence Berkeley Laboratory, University of California, Berkeley, CA 94720

Frederick W. Yeater
Building 80, Room 109, Lawrence Berkeley Laboratory, University of California, Berkeley, CA 94720

Jensen Young
Building 4, Room 203A, Lawrence Berkeley Laboratory, University of California, Berkeley, CA 94720